Science Under Islam
Rise, Decline and Revival

LULU EDITION

Science Under Islam
Rise, Decline and Revival

Sayyed Misbah Deen
Emeritus Professor of Computer Science
University of Keele

LULU EDITION

Publisher: *Lulu.com*
http://www.scienceunderislam.com

Science Under Islam - Rise, Decline and Revival

LULU EDITION
Publisher: *Lulu.com*

ISBN 978-1-84799-942-9

The book can be ordered from *http://www.lulu.com*
See also *http://www.scienceunderislam.com*

© S. M. Deen, 2007
All rights reserved.

No part of this publication may be reproduced, stored in a retrieval system, or transmitted, in any form or by any means – electronic, mechanical, photocopying, recording, or otherwise – without the prior written permission of the author.

*In memory of
my grandfather, a celebrated sufi,*

Sayyed Abdullah Ibn Shahzaman
of Jawar

To
Asli

With Best Wishes

12 June 2010

Acknowledgements

The first person I should mention is my maternal uncle Dr Akhlaq Hussain Ahmed, a Bangladeshi politician, who asked me repeatedly to write this book. I do not think I would have written it without his constant encouragement. Dr Rashid Jayousi of Al-Quds University (Jerusalem) provided me much material, not only from many Arabic sources, but also from his own deep knowledge of the subject area. Another person with a profound knowledge of science and Islam, who provided invaluable help, is Ahmed Al-Shabibi. He read through all the chapters, made important comments and suggestions, and also supplied me with some interesting references.

Two persons who assisted me with the front-cover design are Ryad Soobhany and Kapila Ponnamperuma. I am further indebted to Ryad for reading the whole book thoroughly and for making important observations from the perspective of a young Muslim scientist. And lastly, but most gratefully, I must thank Dr Jonathan Lageard of Manchester Metropolitan University, not only for reading the book thoroughly but also for making invaluable comments and suggestions for improvement. Some of their comments on the book are quoted on the back cover.

Finally, the book was a family endeavour. My wife Rudaina and my daughter-in-law Nivan, both Arabs, helped me with Arabic, another daughter-in-law, Ferina, a medical doctor, assisted in chapter 8 (Medicine), and my two sons, Rami and Alvin, edited and proof-read the book and organised its publication and website. And then Yara, just three, who helped her grandpa by keeping him constantly amused and thus sane, during the insanely hectic publication time.

S. M. Deen, 2007

CONTENTS

Preface   xi
Terms and Arabic Names   xiii

Part I: Historical Background   1
Part II: Rise of Science under Islam   73
Part III: Examination of Success and Decline   115
Part IV: Future Prospect   191

References and Sources are listed at the end of every chapter, but are not shown in the chapter contents given below.

PART I: HISTORICAL BACKGROUND

Chapter 1: Introduction   3
    1.1 Quran and Hadiths on Science and Knowledge
    1.2 The Spirit of the Abbasids
    1.3 Mamun's Dreams
    1.4 Ancient Greek Science
    1.5 The Plight of Science in Ancient Christendom

Chapter 2: History of Early Islam   14
    2.1 The Prophet
    2.2 The Rashidun
    2.3 The Umayyads
    2.4 The Abbasids
    2.5 Spain
    2.6 Shias and Fatimids

CONTENTS

Chapter 3: The Mutazila Movement — 42
    3.1 The Basic Mutazilite Philosophy
    3.2 Doctrine of Khalq-i-Quran
    3.3 Asharite Opposition
    3.4 Fall of the Mutazilites
    3.5 Concluding Remarks

Chapter 4: Hadiths and Sharia — 53
    4.1 Hadiths
    4.2 Sharia Law
    4.3 Limitations and Consequences

PART II: RISE OF SCIENCE UNDER ISLAM

Chapter 5: Mathematics — 75
    5.1 Numbering Systems
    5.2 Arithmetical Operations
    5.3 Algebra
    5.4 Geometry and Trigonometry
    5.5 Cryptography
    5.6 Other Mathematical Works

Chapter 6: Astronomy — 88
    6.1 Spherical Astronomy
    6.2 Planetary Models
    6.3 Astronomical Instruments

Chapter 7: Other Sciences — 102
    7.1 Optics
    7.2 Chemistry
    7.3 Geography
    7.4 Mechanics and Machines

Chapter 8: Medicine — 107
    8.1 The Galenic System
    8.2 Healthcare and Practitioners
    8.3 Hospital Systems
    8.4 Some Medical Giants
    8.5 Beginning of the End

PART III: EXAMINATION OF SUCCESS AND DECLINE

Chapter 9: Causes for Decline   117
    9.1 Social Attitude
    9.2 The Nature of the State
    9.3 Other Causes (Higher education,
                    Collegiality & dissemination,
                    Funding support, and
                    Inadequate Islamic law)
    9.4 Summary and Conclusion

Chapter 10: Greatest "Heretics" and Their Nemesis   144
    10.1 al-Kindi
    10.2 al-Razi
    10.3 al-Farabi
    10.4 Ibn Sina
    10.5 Ibn Rushd
    10.6 Ibn Khaldun
    10.7 al-Gazzali

Chapter 11: Failures of the Later Islamic Empires   165
    11.1 India: The Mughal Empire
    11.2 Turkey: The Ottoman Empire

PART IV: FUTURE PROSPECT

Chapter 12: Some Later Reformers — 193
  12.1 Sayyed Ahmad Khan (India)
  12.2 Syed Ameer Ali (India)
  12.3 Jamaluddin Afghani
  12.4 Muhammad Ali (Egypt)
  12.5 Amanullah Khan (Afghanistan)
  12.6 Reza Khan (Iran)
  12.7 Kemal Ataturk (Turkey)
  Conclusion

Chapter 13: Modern Rethinking on Reform — 213
  13.1 Review of the Early Practices, Sharia and Ulama
  13.2 Unreliability of Hadiths
  13.3 New Reformist Thinking
  13.4 Concluding Remarks

Chapter 14: Challenges for Revival of Science — 232
  Drawing from the whole of the book and exploring the challenges based on the previous chapters, particularly chapter 13.

  14.1 General Environment
  14.2 Review of Rise and Decline
  14.3 Political Reforms for a Democratic Infrastructure
  14.4 Religious Reformation of Islam
  14.5 Socio-religious Education
  14.6 Way Forward Towards Revival
  14.7 Conclusion

Glossary — 251

Index — 259

PREFACE

The book is intended for anyone interested in finding out the extent to which science was developed under Islam, the sublime height it reached, and how and why it declined. It also discusses how the spirit of science can perhaps be rekindled among the modern Muslims, backed with a rational religious attitude, as advocated by many Muslims scholars over the recent centuries. The book is written from the Muslim perspective, with Islamic reformation in mind, ensuring that the criticism, analysis, discussion and remedies presented fall within the scope and culture of Islam. After all Islam is the only major religion whose holy Book has repeatedly asked its adherents to reflect on and to understand God's creation – an unparalleled call to study science. It is argued in this book that the failure to undertake that challenging study is a root cause of the relative backwardness of the Muslim world today, and that the current new thinking among the Muslim scholars shows a way forward.

Let us begin with a quotation from Abdus Salam, the late Professor of Theoretical Physics at Imperial College (London), who was awarded the Nobel Prize for Physics in 1979 (the first Muslim ever to get a Nobel Prize):

> "There is no question but today, of all civilisations on this planet, science is weakest in the lands of Islam. The dangers of this weakness cannot be over-emphasized since the honourable survival of a society depends directly on the strength of its science and technology in the condition of the present age."

From the Foreword in Science and Islam by Pervez Hoodbhoy, Zed Books, London, 1991.

This view of Salam perhaps explains the underlying reason for the current plight of the Muslims all over the world. He was one of my teachers during my PhD study in Physics at Imperial College. In that Foreword, Salam asked to (but did not) see remedies to the current situation. I do not claim to have provided any remedy that would have satisfied my old teacher, but I have tried to put forward some ideas.

Many years ago I heard a story from Salam on the Muslim attitude towards science and technology in the middle ages. The clock in a main mosque in Istanbul broke down, which could be repaired only by a Christian mechanic (by then the Muslims had lost that ability). The question was how could a Christian be allowed to enter a holy mosque (in earlier times there was no such restriction). Then a much needed fatwa was produced: since the entry of a monkey does not defile a mosque and since a Christian is no better than a monkey, he can be allowed to enter the mosque to repair the clock! Salam concluded the story with the comment that this had been the Muslim attitude to science and technology then, and it is still the attitude now. In this book we argue for change.

My own interest on the Muslim attitude to science goes back to my childhood when I started asking why we Muslims did not use calculations to determine the lunar festival-dates, unlike our Hindu neighbours. This intrigued me even further when I discovered that some minority Muslim sects, such as the Ahmadiya, do use predictions. However, the refusal to predict did not become a problem for me personally until recent times, when in the UK we could not arrange the celebration of Eids with our grown up children (as we did not know the dates in advance). This prompted me to investigate the issue of date prediction (see chapters 6 & 9), which led me to science and technology under Islam, and eventually to writing this book. There were of course other contemporary reasons as well, particularly the plight of the Muslims everywhere in the world today, which motivated me to examine our heritage with a view to finding a way forward into the future.

TERMS AND ARABIC NAMES

The term science as we understand it today was often referred to in the older Muslim literature as:

ancient science, secular science, Greek science or exact science,

and later particularly by the Muslim orthodox as:

foreign science, repudiated science, useless science or even *"wisdom mixed with unbelief"*.

Observe that "unbelief" is the English translation of the Quranic word *kufr*, an unbeliever is a *kafir* (a heretic) – the greatest stigma to a Muslim, but awarded liberally by the orthodox to the Muslim scientists. A *kafir* is eternally damned without any hope of salvation.

The *useful science* was *the religious science*, such as to find prayer times; all other sciences were labelled by the orthodox as *useless science* which in their view a good Muslim should not practice. The reader may come across some of these alternative labels in this book and in other Islamic literatures.

The scholars who contributed to philosophy, science and medicine in the Islamic empire included Arabs, non-Arabs (particularly Persians), Muslims and non-Muslims. In the absence of a suitable common term for them, we shall employ the adjective *Arabic* (following other writers) to qualify these scholars, and the term *Arabics* to refer to them as a people, since they all spoke and wrote their work in the Arabic language. Furthermore, it should be noted that science is universal, there is no Muslim science nor Christian science, although there can be a Muslim or a Christian scientist. We shall however apply the term Islamic science (and equally Arabic science) to qualify science that was developed under Islam by these *Arabic* scholars, some of whom were non-Muslims and most of whom non-Arabs. As against this, the term Arab science will be treated as a subset of Arabic science, implying science that was produced by the *Arab* scholars only.

There are many ways of spelling Arabic names in English, for which I have employed the commonly found English spelling particularly in the Muslim writings. I have also used the name Sulaiman, as favoured by the Muslims, in place of Suleyman for the Turkish sultan. However, for some Arabic names such as Muhammad, there are no common English spellings even among the Muslims. In those cases I have chosen my own preferences. A further point is the use of honorific prefix *al* in Arabic surnames, particularly in the case of important people, such as *al*-Kindi or Khalifa *al*-Mamun. I have generally dropped this prefix, as being unnecessary. And finally I have also favoured Muslim terms Khalifa and Khelafa over their anglicised corruption Caliph and Caliphate.

In another name, I have used Chengis, not Ghengis, as the former is regarded as the more correct transliteration in English.

As for dates, I have opted for CE (Common Era), except where qualified with BCE for Before Common Era, or AH for al-Hijra. That is, an unqualified date is a date in CE. Furthermore, in some dates I have employed the letter b for birth, d for death, c for circa (approximate) and ? for an uncertain date, as is done by others.

Quranic Quotations

The English texts of the Quranic ayahs (i.e. verses) quoted in this book have been taken from the translations of the Quran by Muhammad Assad and Abdullah Yusuf Ali, using ayah numbers as given in the translation of Assad.

PART I

HISTORICAL BACKGROUND

CHAPTER 1
INTRODUCTION

Today Islam is often equated with backwardness, and is treated as a creed that cannot face the world moving forward and hence is always looking backward. It is viewed in the West as anti-science and anti-intellectual, a religion that has nothing to offer to "man and his world" in the science-driven 21st century. Yet, it was Islam that carried forward the torch of enlightenment from the Greeks to the middle ages, laying down a foundation from which arose not only the European Renaissance, but also the early modern science with all its associated triumphs – from the Copernican heliocentric planetary system to modern-day hospital practices. So, how is it that the Islam that had created a civilisation when there was so little of anything, has become so sterile when there is so much more of everything. It is a story of rise and decline in the pursuit of intellectual excellence that should be of interest to all human beings, and particularly to Muslims, who have a legitimate craving to recreate their past glory. This book attempts to tell this story in terms of the scientific developments that took place, the reasons for the subsequent intellectual decline that can be discerned and the spirit for revival that can be rekindled.

The Golden Age of Islam began in the mid-eighth century, with the Abbasid takeover of the Khelafa from the Umayyads, when the military conquests for the Islamic empire was replaced by a strong desire to achieve intellectual conquests on behalf of Islam itself – by revealing its greatness and universality in harmony with rationalism. The inspiration came directly from the hadiths such as: *Go even as far as China to seek knowledge*. The knowledge seekers travelled not only to China, but also to India, ancient Mesopotamia, ancient Egypt, and to Greece where they collected the thoughts of Aristotle and other Greek giants. Their approach was based on freethinking and their objective was to reach the frontiers of knowledge from all directions, and to apply the knowledge gathered to define an intellectual universal Islam, as a choice for all men who inhabited this earth. But soon orthodox Muslims, who looked backward rather than forward, struck back eventually succeeding in banishing the foundation of rationalism from Islam and in creating an Islam that is backward, anti-intellectual and anti-science. The conservatives carried with them the largely uneducated masses, by frightening them with the imagery of the eternal hell-fire that would consume them in the afterlife if they did not tow the orthodox line.

In this book we shall examine the rise and decline of science under Islam, and explore some thoughts on the revival of science in Muslim societies, particularly the change in attitude that would be required for such a revival. As regards this chapter, the presentation will include a background of Muslim interest in knowledge and science from religious perspectives, the start of the Golden Age of Islam and the dreams of Khalifa Mamun. Since the Islamic Golden Age was based on Greek science, we shall also present below brief notes on Greek science and its subsequent plight under Christendom, leading to its cultivation under Muslim rule.

1.1 Quran and Hadiths on Science and Knowledge

The Quran and hadiths (sayings of the Prophet) often exhort Muslims to seek knowledge and to ponder the mystery of God's Creation, the universe and its content. About an eighth of the Quran talks about *tafakkur* and *tashkeel*[1], which can be interpreted as science and technology. The entire universe and its content are signs of God's activity which the Muslims are urged to reflect upon and understand. No religion has given a more powerful incentive for the study of science. Science is thus essential for fulfilling the Quranic recommendations. There are also many hadiths exhorting Muslims to seek knowledge. Here are some of them [Tur97]:

1. Go even as far as China to seek knowledge.
2. Scholar's ink is holier than martyr's blood.

---

[1] two words together imply meditation (or thinking about) [i.e. *tafakkur*] on the form and structure [i.e. *tashkeel*] of the God's creation.

3. He who pursues the road to knowledge, God will direct him to the road to paradise.
4. The brightness of a learned man compared to that of a mere worshiper is like that of a full moon compared to that of all the stars.
5. Obtain knowledge; its possessor can distinguish right from wrong, it shows the way to Heaven.
6. Seeking knowledge is required for every Muslim

The early Abbasid scholars believed in these Prophetic endorsements, to seek knowledge to reveal the mystery of the God's Creation and thus to build an intellectually vibrant and flourishing Islam. This they achieved with the help of Arabs, non-Arabs, Muslims and non-Muslims.

1.2 The Spirit of the Abbasids

There were many factors that helped create the Abbasid Golden Age. Under the Khalifa Mansur (754-775 CE), Rashid (786-809 CE) and Mamun (813-833 CE), a major drive for learning was initiated, borrowing ideas from the Persians (with their sense of history and culture), from the Christian Syriacs (with their linguistic versatility), from the Harranians (with their Hellenic heritage), from the Indians (with their ancient lore and mathematics), and above all from the Greeks (with their vast storehouse of ancient knowledge). All ideas were infused to establish a superculture, under the aegis of five Khalifas, those three mentioned above and also two others: Mutasim (833-842 CE) and Wathiq (842-847 CE). The Persians, with their imperial heritage, became the favourite of the Abbasids, as the former excelled not only in administration and finance, but also in all branches of arts, literature, philosophy and science.

The Khalifas created the environment for learning and intellectual excellence. Mansur provided the first spark through his great translation project, which was carried forward by his second son Rashid, with even greater enthusiasm. He established the *Khaznatul Hikmah* (Treasure-house of Wisdom) in Baghdad, later called Baitul Hikmah (the House of Wisdom), under the direction of competent scholars. This House became an important centre where translation of scientific and philosophical works from important foreign sources (Indian, Persian and especially Greek) was undertaken. Material prosperity enabled the people to participate in cultural activities and to conduct the search for knowledge. The elite joined in with the Khalifas, in funding the translation work, to demonstrate their culture and respectability.

Sponsorships of scholars became the norm, debate and discourse became the sign of status. It never happened before – as if all the Greek city states, with their manifold scientific endeavours, had squeezed into the single city of Baghdad. If there was ever a time in human history when seeking truth and knowledge was the most fashionable thing to do, then that was it. Let us quote from Kindi (see chapter 10) to get the spirit of the time:

> "We should not be timid in praising truth and in seeking it from wherever it comes – even if it be from distant races or people different from us."

Kindi, a protégé of Khalifa Mamun, was a leading member of the Mutazilite movement (see chapter 4) that became the advocate of rational thought and pioneer of the intellectual and scientific developments under Islam. However, it was Mamun himself, patron of the Mutazilites, who was the principal driving force. It was he who developed a keen interest in foreign cultures, sending delegations of scholars to Asia Minor and Cyprus to bring back Greek books. He even wrote to the Byzantine emperor for books of Greek learning, who after some hesitation complied. Mamun arranged measurement of the diameter of the earth and sent out groups of scientists to investigate the geographic locations of various events described in the Quran (such as ahl al-Qaf and Alexander the Great). He lavishly rewarded poets, philosophers, scientists, writers, and translators. The general intellectual climate of this time was typified by the scholarly gatherings in the houses of the princes, high officials and wealthy patrons of learning. In these gatherings, the old scholars engaged in heated discussions on philosophical points, while the young scholars developed their speciality in various branches of knowledge. Linguists went on journeys in search of new tongues, geographers undertook expeditions to make new discoveries, librarians toured to find and bring back new books of learning. Many rich families contributed by supporting private endeavours and private libraries.

Within two hundred years of the death of Prophet Muhammad, book industries and libraries were established in every corner of the

Muslim world. In the middle of the 13th century, there were 36 libraries (apart from al-Hikmah) in Baghdad alone. The place abounded with authors, copiers, illuminators, librarians, collectors, buyers and sellers of books. Readers, who could borrow books, came from all classes of the society and all religious affiliations. Some libraries were royal, some public, some private and some specialised. Books sold in one book-store alone had more than sixty thousand titles ranging over many subject categories, including: language and calligraphy, Christian and Jewish scriptures, the Quran and commentaries, histories, government works, court accounts, pre-Islamic and Islamic poetry, Muslim schools of thoughts, biographies of the men of learning, Greek and Islamic philosophy, mathematics, astronomy, medicine, literature, popular fictions, literature, travel (e.g. to India, China and Indochina), magic and fables. Although compiled later, the fables of *The Thousand and One Nights* (also called *The Arabian Nights*) depict that period.

Many scholars themselves, such as Kindi, funded intellectual works (of philosophers, scientists, translators, doctors and poets). Most of the translators of Greek books were Christians but most of the philosophers were Muslims. Some doctors were Christians but many more were Muslims. The search of knowledge under Islam was conducted together by Muslims alongside Christians, Jews, Hindus and Zoroastrians. This was unparalleled in the history of any other religion, and has only been equalled in the modern secular time. If you were of a great family, you had to go for scholarship. That was the social norm.

1.3 Mamun's Dreams

Khalifa Mamun was the greatest sponsor of knowledge-seekers. There are two versions of a dream in which he met Aristotle who inspired him. Both are described below, as they are interesting stories to read, the first from [Sab76/p181], and the second from [Fak83/p55-76]:

Dream: Version 1

It is said that one day Khalifa Mamun, dreamed that a bald-headed man with a long beard, fair and ruddy complexion, broad forehead and joined-up eyebrows was sitting on a couch. In awe and with a trembling voice the Commander of the Faithful (i.e. the Khalifa) inquired: Who are you? "I am Aristotle" was the reply. Mamun then asked four questions:

1. What is good?
   "That which is good by reason" was the reply.

2. What's next?
   "Whatever is good by the law".

3. What's next?
   "That which is considered good by the people".

4. And next?
   "Nothing next".

Finally Aristotle advised Mamun to treat like gold whoever advised him about gold and to hold onto the doctrine of *Tauhid* (strict unity of God in Islam). It was the consequence of this dream, so the story goes, that Mamun decided to seek knowledge and for that purpose to accelerate the translations of the books of the ancient philosophers into Arabic.

Dream: Version 2

It is said that Khalifa Mamun spent a restless night, brooding over his objective to establish an era of enlightenment in Islamic thinking through the extension of the Baitul Hikmah (House of Wisdom), initiated by his father Rashid. The extension was nearly finished, with a library, a translation bureau and a school. As his project was about to be completed, he had nagging doubts which kept him awake the whole night. The questions that kept him sleepless were: Was his objective truly worthy in the eyes of God? Was it right to stimulate and spread new ideas? Was it right to revive Greek philosophy – the wisdom of the ancients? Should he encourage reason and logic for a deeper understanding of God's universe and His truth in the revelations sent through the Prophet? But suppose reason and logic led elsewhere away from God, what then? If man could not prove the mysteries in the revelations, would it mean disapproval of God by the intellect of man? What then? The troubled Khalifa finally dozed off. He lapsed into a curious but relaxing dream, in which a smiling long-bearded old man in Greek attire appeared. He recognised Aristotle who gently touched his forehead with enlightenment: "There is no conflict – reason and religion are allies, not enemies", declared the revered sage. Then the sage disappeared. The Khalifa woke up abruptly, relaxed and brimming with renewed confidence in his noble objective to serve God through knowledge, reason and logic.

For Khalifa Mamun to dream of Aristotle even in a story, he had to know of Aristotle and of the Greek contributions to civilisation. It is time we provided that background.

1.4 Ancient Greek Science

The origin of Greek science is usually traced back to Thales (c 624-565 BCE), born in the Ionian city of Miletus[2]. Son of a Phoenician mother, he was involved both in politics and business. His work later took him to both Mesopotamia and Egypt, the two main sources of Greek science in which Ionians played a major early role. He learnt about astronomy and eclipses from Mesopotamian temples, and by sheer luck he correctly predicted the solar eclipse of 585 BCE in Miletus. This aroused a great interest in the whole of Greece, in the systematic observation of nature, with a view to deriving benefit from it. Thales himself, a geometrician, generalised Egyptian calculations on individual right-angled triangles that the square of the longest side in each case is equal to the sum of the squares of the other two sides. The Egyptians knew it to be true for only specific numerical examples. Thales was followed by a succession of Asiatic Greek scientists, mostly from Miletus. Among them was Thales' pupil Anaximander (611-547 BCE) who constructed the first maps, and introduced the sun-dial, both based on ideas borrowed from Mesopotamia and Egypt.

These Ionians continued to visit foreign lands in order to bring back, consolidate and generalise the knowledge gained, thereby contributing to the economy and business, and also to create what they called the *Religion of Science*. Some of the later thinkers devoted

---

[2] Ionia in antiquity was the central Western coast of Asia Minor (present day Turkey) and the adjacent islands (e.g. Samos), settled by the Greeks in around 1000 BCE. Miletus, also founded in around 1000 BCE, was an important costal city in Ionia, but largely destroyed in 494 BCE by the Persians.

themselves entirely to philosophy, one of them was Heracleitus (540-475 BCE), who gave us the statement: "everything is in a state of flux".

The first scientist to fall foul of the orthodox was the Ionian Anaxagoras (488-428 BCE), who moved to liberal Athens in 464 BCE and subsequently became known as the father of Athenian science. He gave scientific accounts of eclipses, meteors and rainbows. He declared that the sun was a vast mass of incandescent metal, and that moonlight was reflected sunlight. Although he successfully defended himself against the charge of impiety, he felt it prudent to spend the rest of his life in his native Asia Minor. Another scientist Aristarchus (c 310-230 BCE) of Samos (an Ionian Island) was the first person to propose the concept of heliocentric planetary system, but he was persecuted for this impious view. In more recent times he was given the title *Copernicus of Antiquity*.

Plato (427-347 BCE) differed much in his philosophical approach from that of his great master and mentor Socrates (470-399 BCE). Aristotle (384-322 BCE), pupil of Plato, deviated from both of his celebrated predecessors, taking much from Anaxagoras. Aristotle was a son of a court physician, and later became the tutor of Alexander the Great. He was the Greek genius who excelled in every branch of knowledge, including logic, ethics, politics, metaphysics, physics, biology, psychology, poetry and rhetoric. He wrote over 400 books on these topics. There are not enough pages in this book to do justice to this extraordinary man. It seems he even thought of the Calculus, beating both the bickering Leibniz and Newton by some two thousand years! This Greek genius was honoured by the early Abbasids, but was detested by the later Muslim orthodoxy – a sure sign of their intellectual bankruptcy.

Democritus (c 470-400 BCE) of Miletus was a contemporary of Socrates and gave us the notion of the indivisible atom as the ultimate constituent of matter, a concept that was revived by John Dalton in the early 19th century, though not experimentally verified until Lord Rutherford determined its physical structure in 1911. However, today we know it to be divisible into other subatomic particles.

Hippocrates (460-377 BCE) of the Greek island of Cos (or Kos, off the South Western coast of modern Turkey) came from a family of physicians, who founded a school of medicine in which he transformed the then medical tradition into a scientific procedure, using fact-based inductive techniques. He insisted that diseases must be treated not as a divine punishment, but as something that has earthly causes to be cured by the physicians. This was a revolutionary new approach, replacing the religious brews and chants by actual medicine[3]. Galen of Pergamon (129-199 CE) developed a medical system, called the Galenic system (see chapter 8) in which he applied ideas from both Hippocrates and Aristotle.

1.5 The Plight of Science in Ancient Christendom

In this section, we shall describe briefly how the orthodoxy in the ancient Christian empires forced their scientists and philosophers to take refuge in the Middle East, and thus paved the way for the cultivation of science under Islam.

In 330 CE emperor Constantine (285?-337CE) moved the capital of the Roman Empire to Byzantium, later named Constantinople (present day Istanbul). Following the Council of Nicaea in 325 CE, Christianity became more doctrinaire, and some 50 years later emperor Theodosius (347-95 CE) turned his empire into an orthodox Christian state. From the third century onwards there were few Christian scientists of real originality in the empire, except in some minority sects such as Monophysites and Nestorians. The Monophysites mainly believed in a single nature of Christ as the divine Word, while perhaps as over-reaction to the Monophysites, the Nestorians in the main believed not only the dual nature of Christ but also viewed him as two persons: a true man and God, and therefore Mary is the mother of a man, not the Mother of God (a complicated concept). However, the final death knell for science and philosophy was issued by emperor Justinian (527-65 CE) who first persecuted and then banished the Nestorians and pagan philosophers from Athens. The great Athenian Academy (a centre of learning) was closed.

On a parallel track in the successive centuries, following the death of Alexander the Great in 323 BCE, many Greek kingdoms, great and small, had been established all over Persia, among them was the kingdom of Bactria (in the present Iranian province of Khuzistan) created by a Greek Governor Diodatus in the third century BCE. However, in 226 CE, Persian emperor Ardashir drove the

---

[3] However, the famous Hippocratic Oath taken by all modern medical students (and also by the 9th century Arabic students – see chapter 8) was not probably written by Hippocrates himself.

Greeks out of Persia and established the secular Sasanid dynasty which lasted until it was defeated by the Muslims in the seventh century.

Jundishapur[4] was a city near the ancient Greek capital of Susa in Bactria. In 489 CE, anticipating troubles from the Roman Christian orthodoxy, the Nestorians wisely transferred their scientific centre from Edessa (now Urfa in South East Turkey) to a place close to Jundishapur, which was then part of Khuzistan under the secular Persian rule. When emperor Justinian closed their Athenian Academy and banished the Nestorians from Athens in 529 CE, the Nestorians scholars left the Roman Christian Empire to avoid persecution. Some of them went to establish a centre of learning (Academy) in Jundishapur, then under the reign of the Sasanid ruler Anushirwan (531-579 CE), where they continued the study of philosophy, science and medicine, alongside pagan philosophers. The area was conquered by the Muslims in 636 CE, with its great Academy and its medical treatment centre intact. As we shall see later, this Academy became the catalyst for the Golden Age of Islam (see also chapter 8 on medicine)

In Jundishapur, the Nestorians continued to carry out important works in medicine, astronomy and mathematics, and they translated their instructions into Syriac[5] to assist the local recruits. There were also the Monophysites, who were also persecuted by the orthodox Christians; they were working in Syria on similar subjects and translated important scientific and philosophical works into Syriac. A third group that assisted the Muslims in such translations were the Sabians of Harran in Mesopotamia. Apart from these Greek sources, ideas have also been absorbed by the Arabic scholars from the Iranian Pahlavi (the ancient language of Iran) and Indian Sanskrit literature, but they are less well-documented (see pages 75-76).

By 644 CE, the Arabs had conquered Syria, Iraq, Iran and Egypt, and thus inherited the Greek centres of learning in those countries. The translations and scientific work continued in Syria and Khuzistan during the Umayyad period (661-750 CE) with its capital in Damascus, but mostly independently of the rulers. The

---

[4] Strictly Jundi-shapur, which means a beautiful garden, but it is also spelled as Jundeshapur, or even Gondeshapur in some English texts.

[5] The Semitic language (based on the dialect of Aramaic) of ancient Edessa, where the scholars at the great Christian centre of learning wrote their works in this language between the 3rd and 7th century CE.

Umayyads were too busy with expanding the empire and crushing internal revolts. The exception to this was Khalid bin Yazid (a son of the second Umayyad Khalifa), who was known as *the wise man* for his knowledge of astronomy and chemistry. He is said to have got translated a number of Greek scientific works into Arabic for his personal use. However the Golden Age started with the Abbasids.

References and Sources

[Fak83] M. Fakhry: "Philosophy and History", The Genius of Arab Civilization, J. R. Hayes (Ed), Second edition MIT Press, 1983, pp55-76.

[Sab 76] A. I. Sabra: Ch 7 - The Scientific Enterprise, Islam in the Arab World, ed. B. Lewis, published by Alfred A. Knopf, New York, 1976, pp.181-200. Abdulhamid I Sabra is a great Egyptian scholar and an authority in Arabic science.

[Sart27] George Sarton: Introduction to the History of Science, Carnegie Institute of Washington, 1927. He has three huge volumes (I, II, III) on science from Homer to the 14th Century – the size is equivalent to some 5500 A4 pages. Most scientists (including Arabic, Chinese, Indian and Japanese) are individually listed and contributions described. These are indeed the books that made Arabic contributions to science well-known.

[Sart53] George Sarton: A History of Science, published by the Oxford University Press 1953 on ancient science, containing some 1000 A4 pages.

[Sart59] George Sarton: A History of Science, Harvard University Press, 1959. This book covers Greek Science in the last three centuries BCE. The size is equivalent to some 1000 A4 pages.

[Tur97] Howard R. Turner: Science in Medieval Islam, University of Texas Press, 1997, ISBN 029278 1490.

CHAPTER 2
HISTORY OF EARLY ISLAM

In order to provide a background to the development of science under Islam, we shall present here the early history of Islam, tracing its rise from the birth of Prophet Muhammad until the destruction of Baghdad in 1258 CE by the Mongols. We shall also include in this presentation, parallel events in Iran, Egypt and Spain, but excluding the Crusades which did not have much relevance to the development of Islamic science. The histories of both Egypt and Iran will be outlined right up to modern times for ease of later references. The Indian Mughal and Turkish Ottoman empires will be examined in chapter 11 in Part III of the book.

2.1 The Prophet

The city of Mecca was a major centre of pilgrimage and trade in Arabia. Its dominant tribe, the Quraish, claimed descent from Abraham through his son Ismail. The father and son were (and are) believed to have created the cubical (Kaaba) temple as the house of God in Mecca. The area of some 20 miles radius around the Kaaba was called the Sanctuary, within which no fighting was allowed. The Kaaba, filled with idols by the polytheist Quraish, became the most important place of worship in the whole of Arabia, with pilgrims flocking to Mecca, particularly during a period of three months every year when fighting among the Arab tribes was forbidden. The supervision and control of the Kaaba gave the Quraish a special prestige and provided the main source of their wealth.

Muhammad, the Prophet of Islam, was born in Mecca in 570 (or 571) CE as a posthumous child of Abdullah Ibn Abd al-Muttalib of Mecca and Amina bint Wahb of Medina, both of the tribe of Quraish. His mother Amina died when Muhammad was six, and his grandfather Abd al-Muttalib died when he was eight. The task of bringing him up then fell on his uncle Abu Talib, the full brother of his father and the new chief of the clan of Hashim, the father of Abd al-Muttalib.

Hashim, the great grandfather of the Prophet, was a leading Quraish aristocrat of a previous generation, and because of his high status his descendants came to be known as the Hashemites. His son Abd al-Muttalib was also a leading member of the Meccan aristocracy, although his fortune seems to have declined later. It would appear that his son Abu Talib, the guardian of young Muhammad, did fall into hard times, with diminishing influence, but was still very highly respected. Wahb, the maternal grandfather of the Prophet, was the leader of a small branch of the Quraish living in Medina, but he died before the marriage of his daughter Amina.

From his childhood Muhammad was known to have been very contemplative. In due course, he joined his uncle Abu Talib in his business trips, which took Muhammad to other parts of Arabia and also to Syria. He later worked for, and married, a rich widow called Khadija, who gave him four daughters, the only children of Muhammad to reach adulthood.

As part of his contemplation, Muhammad often meditated for days and nights in a cave called Hira near Mecca. One night, he had a vision, the vision of Archangel Jibra'il (Gabriel), who commanded him to read:

"Read in the name of thy Sustainer, who has created –
Created man out of a germ-cell.
Read, for thy Sustainer is the most bountiful One,
Who taught man the use of pen –
Taught man what he did not know. "
                              Quran [96:1-4].

Muhammad first protested that, being unlettered, he could not read, and then finally recited those verses with the Archangel. Observe that these very first verses sent down for the Quran, talked about reading, writing (pen) and knowledge. And yet the Muslim society later became very anti-knowledge in science and philosophy, as we shall discover in this book.

After the vision that evening, Muhammad came home to Khadija, still shaken and frightened by the encounter. Khadija reassured him and calmed him down. He was forty, and started preaching immediately. He declared that there was only one God to worship, and he (Muhammad) a human being was His messenger. A Muslim must surrender to the will of God, pray to Him, keep fast, give to charity, obey parents, care for the needy and orphans, do good, not make excessive profits (in business), and spend the wealth created to do good. These messages did not find favour with the Quraish, who were selfish and greedy, without any social conscience (in today's terms) or any sense of justice or fairness. For them wealth and power were the only things that counted in life.

During the next 23 years the Quran was revealed by God to the Prophet, the revelation completing just before his death. Khadija became the first Muslim. The next was either his friend Abu Bakr or, according to some sources, his cousin Ali (son of Abu Talib) whom he (Muhammad) took earlier as a ward. Soon afterwards, Uthman Ibn Affan, a member of the richer Umayyad clan (Umayya was a first cousin of Abd al-Muttalib, through the common grandfather Abd al-Manaf) became Muslim, and next the virulently anti-Muslim Umar Ibn Khattab succumbed to the message of Islam. These four later became the first four Khalifas, called *rashidun* (the rightly guided ones).

We shall digress here a little to mention the origins of the two dynasties that followed these four rightly guided Khalifas. Muawiya, a cousin of Uthman and son of the last anti-Muslim Meccan leader Abu Sufian, made himself the fifth Khalifa by might and established the Umayyad dynasty. The descendants of Abbas, an uncle of the Prophet, created the Abbasid dynasty after annihilating the Umayyads, justifying their legitimacy to rule on the basis of his kinship to the Prophet. And yet Abbas himself was not an early convert – in fact he took arms against the Prophet at the first battle (i.e. Badr in 624 CE) between the Prophet and the Quraish. He was captured in the battle but was released unconditionally by the Prophet, as he was his uncle. Nevertheless, Abbas did not become Muslim until much later (see below). Once converted to Islam, he became an ardent supporter of Ali, his nephew and the fourth Khalifa.

Returning to the main narrative, Muhammad was often persecuted by some of the Quraish leaders in Mecca. Being incensed by one particular instance of persecution, his uncle Hamza entered Islam, and from then on he loyally served the Prophet, until he was killed at the battle of Uhud in 625 CE (see also below). However, despite the support of Hamza, life in Mecca became perilous for Muhammad and his followers, as the Quraish did not like this new religion of uncompromising one God, which threatened their pilgrimage income. They all worshiped idols, some 360 of them (representing a range of gods) were housed in the Kaaba, the central attraction for the pilgrims. However, Islam continued to spread with more people becoming Muslims, which in turn increased the hostility of the Quraish.

The death of Khadija in 619 CE devastated Muhammad. Shortly afterwards Abu Talib, who did not become a Muslim but nevertheless protected his nephew, also passed away, leaving Muhammad vulnerable to physical attack by the Quraish. That year was later referred to as *the year of sadness*. Emboldened by the death of Abu Talib, the Quraish increased their persecution and harassment of Muhammad and Muslims, and eventually decided to eliminate him. Being aware of such a possible danger, Muhammad had already negotiated with his new converts in Medina[1], a plan for the migration (*hijra*) of all Meccan Muslims there. He himself left Mecca, just as the assassins were about to surround his house. Having missed him in his house, they searched for him on the road to Medina, but the Prophet was able to evade them and arrived in Medina a few days later to a great welcome. This was the year 622 CE, the start of the Islamic lunar year al-Hijra (AH). In Medina, he established a Muslim society (an *umma*), but also had to fight three battles against the Quraish: Badr (624 CE), Uhud (625 CE) and finally the Trench (627 CE).

Following the failure of the Quraish at the battle of the Trench, Abu Sufian the shrewd Meccan leader saw the writing on the wall and quickly realised that the Quraish would lose all prestige and authority among the Arab tribes, unless they came to an accommodation with Muhammad. A year later (628 CE) the Meccans were forced to sign a peace treaty (Treaty of Hudaibiya) with the Prophet. Following this treaty, the Prophet performed the hajj in Mecca the year after, when his uncle Abbas and many others embraced Islam.

After the Quraish broke some terms of the Treaty, the Muslim army entered Mecca unopposed in 630 CE, when most Meccans, including Abu Sufian, became Muslims. Muhammad removed all the

---

[1] The ancient name of Medina was Yathrib, but renamed later as the Medina (City) [of the Prophet].

360 idols from the Kaaba and declared it as the House of the one God of Islam. The Prophet was magnanimous in victory – he took no revenge against the people of Mecca. Controlling his anger, he even forgave Hind, the notorious wife of Abu Sufian, who at the battle of Uhud chewed the liver of the fallen Hamza (Prophet's dearest uncle, brother of Abbas) and made ornaments from his body-parts. After a few days, the Prophet returned to Medina, the Prophet's City, which soon became the centre of power for the whole of Arabia, since most tribes now accepted Islam. In the spring of 632 CE, he visited Mecca to perform his last hajj, often referred to as the Farewell Hajj, in which the universal message of Islam and the equality of all men in the eyes of God irrespective of race or colour were powerfully re-stated.

The Prophet of Islam was an unassuming and humble man, kind and compassionate, leaving everything to the will of God. He repaired his own clothes, mended his own sandals, and most of what he got he gave away to charity, often starving himself. Once when his favourite daughter Fatima came to him seeking material assistance, he showed her his own empty stomach, implying that he himself had nothing to eat. He always claimed to be an ordinary man with an extraordinary duty, as God had chosen him as the messenger to receive and spread the message of the Quran. In return for the message the people are asked to love each other:

> "Say (Oh Prophet): No reward do I ask of you for this (message) other than (that you should) love your fellow men."
> Quran [42: 29].

The Prophet warned his people *not* to follow him on non-religious matters. When his suggestion on how to reap a better date-crop proved wrong, the embarrassed Prophet asked his followers to heed only his religious advice. He also told his followers that God would not help them unless they helped themselves, just praying was not good enough. He asked them to seek knowledge from everywhere, to go even to China if needed. That was the basis on which Islam was founded by Muhammad. At his lowest point in Mecca (when his life was threatened), the Prophet of Islam never compromised his faith, and at his highest point when Mecca was conquered, he showed unbelievable magnanimity – the sign of a truly great man. In ten years he created a state out of discordant and fratricidal Arab tribes, who would later take on successfully the might of the Persian and Roman empires.

After the death of Khadija but before the Hijra, Muhammad was persuaded to marry Ayesha the young daughter of Abu Bakr. In Medina he married Ali to his youngest and favourite daughter Fatima, the only daughter to survive Muhammad. Uthman married two of his other daughters, first Ruqaiya and then after her death Umm Kulthum, while the eldest daughter Zainab married a nephew of Khadija.

During this period, fighting created many Muslim widows, and Muhammad urged his followers to take them as wives, so that they could be looked after. He himself set an example by marrying Hafsa, the widowed daughter of Umar, to support her. Hafsa, like her father, was highly accomplished, able to read and write; she soon became a close friend of Ayesha, but the Prophet's household was not entirely peaceful. He had many wives who formed rival groups with inevitable tensions amongst them, none more serious than the antagonism between Ayesha the favourite wife of the Prophet and Fatima his favourite daughter. This was compounded by frictions between Ayesha and Ali, which later spilled over into the political arena. However, by 631 CE, only Fatima out of all his children was alive; she with Ali and her two sons Hasan and Hussain constituted his immediate family (*ahl al-bayt* – the People of the House [of the Prophet]); Abu Bakr, Umar and Uthman (all related to the Prophet by marriage) formed the next closest tier. Abbas, a late convert, became a close confidant of Ali, his nephew.

Shortly after his return to Medina from the hajj in 632 CE, the Prophet fell ill and died within a fortnight on 8th June, without appointing any successor. Umar immediately saw the danger of disintegration of the embryonic Islamic state and quickly negotiated a deal with the Medinan tribal leaders to declare Abu Bakr as the successor (Khalifa) to the Prophet. Ali was initially reluctant, but later accepted the deal, apparently only after the death of Fatima, who was opposed to Abu Bakr, father of Ayesha.

2.2 The Rashidun

At the death of the Prophet, most Arab tribes presumed their agreements with the Islamic state to be over and therefore decided to return to their old ways. Revolts and war broke out everywhere, creating a serious challenge to the rule of Abu Bakr, but he succeeded in crushing them and thus creating an Islamic state where allegiance lay to the state rather than to the individual (as the head of the state). Abu Bakr died within two years in 634 CE, but before his death he declared Umar to be his successor.

Many of the rules and regulations of Islam (including the Islamic year al-Hijra) were founded under Umar, who held power for 12 years and expanded the Islamic empire to include Iraq, Syria, Palestine and Egypt – former domains of the Sasanid and Byzantine empires. In 644 CE, a disgruntled Persian prisoner-of-war attempted to kill Umar. Fatally injured, he appointed a six-man committee to choose a successor. They narrowed the choice between Ali and Uthman. Uthman was eventually selected, as he agreed to abide by the precedents of the two earlier Khalifas (Abu Bakr and Umar), to which Ali refused to commit.

However, the choice of Uthman as Khalifa was a curious one, given that he was nearly seventy, known to be inept, work-shy and undistinguished as a leader during times at war or in peace. He was however, an early Muslim, pious, compassionate, generous and related to the Prophet. As Khalifa, he carried out extensive public works, improved the administration and finance, and replaced the austerity of Umar with general affluence. His Achilles' heel was his kinsmen from the power-hungry Umayyad clan and his blindness towards their misdeeds. He surrounded himself with these kinsmen, who dominated his rule. He appointed the unscrupulous and ambitious Marwan (a cousin) as his vizier, Muawiya (another cousin and son of Abu Sufian) the governor of Syria, another kinsman governor of Egypt, and other relatives to similar high positions. He distributed large sums of money from the state treasury to his kinsmen, while the soldiers were not paid. He responded to the complaints by lashing out at the critics for their "ungratefulness" in the face of his own kindness and generosity, *vis-à-vis* the harshness of Umar (see also chapter 13, footnote 1).

To pacify the unrest in Egypt, he appointed Muhammad Ibn Abu Bakr (brother of Ayesha and adopted son of Ali) as the new governor of Egypt, to replace his kinsman. On their journey to Egypt, Muhammad and his party caught a messenger from the Khalifa carrying a secret letter (with the Seal of the Khalifa) to the current governor ordering him to kill Muhammad immediately. Later, the letter was believed to have been forged by vizier Marwan, unbeknown to the

Khalifa. The disgruntled soldiers and Muhammad's supporters rioted in Medina, during which time, according to some reports, Ali tried in vain to negotiate a compromise. When the Khalifa Uthman refused to hand over Marwan, the soldiers went berserk, broke into his house and one of them killed him. Although Muhammad was not the killer, the Umayyads later accused him of the murder. Marwan and his ten year-old son Abd al-Malik (both later Umayyad Khalifas), who were hiding in Uthman's house, managed to escape.

At Uthman's death, Ali was proclaimed as the fourth Khalifa in 656 CE, which Ayesha opposed. Ayesha previously rebelled against Uthman, but the realpolitik was such that this time she joined forces with Muawiya to blame Ali for involvement in the assassination of Uthman, as he (Ali) refused to handover the accused murderers to the Umayyads for tribal punishment. Ayesha led an army (signs of early egalitarianism in Islam, which disappeared later) against Ali, but lost and retired later a rich widow, partly due to the compensation she received for her property in which the Prophet was buried. In contrast, Fatima asked Khalifa Abu Bakr twice for the inheritance of her father, but he refused on grounds that 'the Prophet has no heirs'. Apparently Abu Bakr did not believe her assertion that her father had promised her his legacy from war booty before his death. Fatima died soon afterwards in 11 AH. It may be noted that the Islamic state was more austere at that time and it became prosperous only later during the time of Umar[2]. Ayesha, who was 18 when the Prophet died, subsequently became the respected source of many hadiths, and died at the age of 64 in 678 CE. Hafsa, her friend, became the Custodian of the copy of the Quran (see below).

Ali moved his capital to Kufa, but Muawiya continued to incite the populace (displaying the blood-stained clothes of Uthman) accusing Ali of complicity. When Ali dismissed him as the governor of Syria, Muawiya declared himself as the Khalifa in Jerusalem. There were several battles between Ali and Muawiya, but the results were indecisive. Two major new Muslim groups came into existence: one

---

[2] Umar gave annual stipends to the war veterans and the companions, cousins and widows of the Prophet, the amount varying from 200 dirhams for undistinguished ordinary soldiers to 5,000 dirhams for the companions and cousins of the Prophet, with 10,000 dirhams reserved for the widows of the Prophet, except for Ayesha, who received 12,000 dirhams. The living cost of a labourer was about 500 dirhams a year. The stipends were increased later by Khalifa Uthman, and were continued to be paid to the descendants by the Umayyads [Rog06].

called *Shi'at al Ali* (party of Ali – the forerunner of Shia'ism) which supported Ali, and the other called *Kharijis* (*seceders* – literally "those who went out") which developed extremist ideas. The Kharijis decided to return Islam to more orthodoxy. They blamed Ali for not being tough enough with Muawiya, and then decided to kill them both as the enemies of Islam. They succeeded in murdering Ali in 661 CE in Kufa, a stronghold of Ali. Thus the Kharijis killed the revered fourth Khalifa of Islam, in the name of religion.

The Kharijis also created the abominable concept of *takfir* (denunciation) to denounce any Muslim who did not agree with them as a *kafir* (unbeliever) who could be killed. The idea of *takfir* goes against the very grain of Islam, since in Islam only God can judge who is or is not a true Muslim, and hence no Muslim can excommunicate another. And yet *takfir* became a tool in the hands of the later orthodoxy to condemn scholars (including scientists) as infidels and then to execute some of them, as we shall see in this book. Today *takfir* is used by some Muslim extremist groups to justify killing *any* Muslim who does not subscribe to their extremist version of Islam.

Returning to the main narrative, those four Khalifas were known as the *Rashidun* – the rightly guided ones, who used the state to advance the cause of Islam and Muslims, while the later Khalifas employed the state to serve themselves and their egos. While the first four lived simple lives with meagre sustenance from the state in austere conditions (except for Uthman who had previous personal wealth), the later ones lived in palaces in opulence and luxury, paid for by the state. The first four Khalifas were addressed directly by their names. Umar thought the title *Khalifa to the Prophet* as too grandiose for him, and therefore called himself *Khalifa to the Khalifa to the Prophet*, and accepted what he considered to be a humbler title *Amirul Mu'mineen* (Commander of the Faithful). The reverential attachments, such as *Radiallahu Anhu* or *Hadrat* were later innovations.

The Islamic state that evolved under the Rashidun was a volatile one. While the foundation laid down by the Prophet and the Rashidun remained, the structure changed through the struggle for power. As the lands of Islam spread, so did the dissensions and conflicts among the rulers, among the ruled and among all people. As the people of different nationalities and culture swelled the Muslim population, so did the requirement to reinterpret the holy scripture and to re-examine the traditions for new laws for all Muslims; but the egalitarianism conceived by the Prophet was never realised, in fact it moved further and further away. However, one thing Muslims possessed and carefully protected was and is the Quran, as the unquestionable source of divine guidance. Here is how it was preserved.

Story of the Quran

The first compilation of the Quran was initiated by Abu Bakr at the advice of Umar, who was concerned at the death of so many Muslims who knew the Quran by heart. The task of collation was given to Zaid bin Thabit, who had been the Prophet's secretary and scribe. After the death of Abu Bakr, Umar completed the work, and handed the compiled Quran to his well-educated daughter Hafsa (the Prophet's wife) for safe keeping. However, the Quran was the first book in Arabic, but the Arabic alphabets did not have any vowel symbols at that time. This led to variations in reading forms, which was removed by Uthman, with a copy in the dialect of the Quraish, in which seven different variations in pronunciation were allowed. This copy was also checked by the same Zaid bin Thabit, in consultation with the leading surviving Companions of the Prophet, including of course Ali. Uthman ordered the destruction of all other copies and versions throughout the Khelafa. However, there were still some inadequacies in the Arabic script, which continued to evolve. The Umayyad Khalifa Abd al-Malik (685-705) produced a new written form of the Quran, with the latest version of the script that was then available. There were further script changes in the Arabic language, which became settled towards the end of the 9th century, when the final written form of the Quran was set down. Its content however remains the same as finalised by Uthman.

2.3 The Umayyads

After the death of Ali in 661 CE, Muawiya became the unchallenged Khalifa, even though Hasan the eldest son of Ali and Fatima initially opposed him. It is interesting to read the correspondence between Hasan and Muawiya, Hasan claiming the Khelafa by the right of his *family* and Muawiya by the right of his *experience*. Hasan's letters also showed how upset Ali had been on being denied three times his legitimate right (as he saw it) to be the Khalifa. But at the end, *ex-*

perience won, with Muawiya accepting Hasan as his successor, a promise that satisfied Hasan but was not intended to be kept by Muawiya. The new Khalifa moved the capital of the Islamic state from Medina to Damascus, where his power was based on the loyalties of the tribal chiefs, whom he kept happy. When Hasan died prematurely at around the age of forty in 669 CE (allegedly poisoned by Muawiya's agents), his younger brother Hussain took up the cause. But Muawiya, who by then had become quite strong, declared his son Yazid to be the next Khalifa.

The Shias regard Ali as the first Imam, and then Hasan, followed by Hussain. The principal difference between the Shias and Sunnis is that the Shias do not accept the deeds and sayings of the first three Khalifas as valid precedents. According to them, Ali belonging to the family of the Prophet (*ahl al-bayt*) should have been the first Khalifa. In addition they believe in the concept of Imam, who is given a wider liberty in interpreting Islam. The only hadiths the Shias pay attention to are those that came through their Imams (see later). While the term Shia comes from *Shiat al-Ali*, as stated earlier, the term Sunni comes from *ahl al-sunna* (the people of the tradition [of the Prophet]).

After Muawiya's death in 680 CE, his son Yazid became the next Khalifa, but Hussain refused to acknowledge his legitimacy. Despite his father's last advice to leave Hussain alone, Yazid began hostilities against Hussain who then sought sanctuary in Mecca. Yazid responded by despatching agents to Mecca to assassinate him, violating the sanctity of the holy city. At this, Hussain decided to take a stand against Yazid. He left Mecca, ignoring advice, in 680 CE with his family members and some 50 supporters, for Kufa where further support was expected. At Kerbala on the bank of the river Euphrates, his small band was surrounded by the army of Yazid commanded by Umar bin Saa'd, despatched by Obaid Allah Ibn Ziyad, governor of Kufa, an ardent former supporter of Ali. On 10th Muharram (the first month of the Muslim lunar calendar), Yazid's army slaughtered them in cold-blood, except for the 22 year-old Ali (Zain al-Abedin), son of Hussain, who was later found seriously ill in a corner of a tent[3]. The other sons of Hussain, including his young

---

[3] Most accounts describe Ali (Zain al-Abedin) as a sick child in the tent. In fact he was 22, already father of 2½ year-old Baqir. Shimar wanted to kill him too, but

baby Ali Asghar were killed. The last to die was Hussain himself, his head was severed from his body by Sinan Ibn Anas Ibn Amar al-Nakhlai (an Umayyad soldier, under Shimar who led that attack), defiling his corpse (a crime in Islam). Yazid promised extra money to the soldiers. The head was taken to Damascus for showing it to Yazid. While Hussain's body was buried at Kerbala, there is dispute as to where the head was buried[4]. The date of this slaughter was 10 Muharram in 680 CE, a date that is commemorated by all Muslims, most prominently by the Shias, every year as *Ashura* (from the Arabic word *a'shara* meaning ten).

Imam Ali Zain al Abedin retired from politics accepting the de facto rule of the Umayyads and devoting himself to religious duties. His mother was Shahr Banu, the daughter of the last Persian emperor Yazdigard III, and hence he had 50% Persian blood, non-Arab blood becoming very common in many later Imams, as well. Before his death in 712 (or 713) CE, he nominated his eldest son Muhammad Baqir as his successor, the 5th Imam. Baqir's son Jafar Sadiq, the 6th Imam, also opted for a non-political life in Medina, dedicating himself to scholarship and devotion, but was nevertheless, according to the Shia sources, killed by Khalifa Mansur with poison in 765 CE (see section 2.4).

It is ironic that Muawiya claimed the legitimacy of his Khelafa by linking himself to the Prophet through Abd al-Manaf, the common grandfather of Abd al-Muttalib and Umayya. The wholescale slaughter of the *ahl al-bayt* at Kerbala made it harder for the Umayyads to claim this thread of legitimacy, and soon became the rallying call for anti-Umayyad revolts.

Understandably this massacre sent a shiver throughout the Muslim world. Yazid was declared *fasiq* (transgressor) and was confronted with retaliatory civil wars on several fronts, but he attacked Mecca and Medina, and set fire to the Kaaba, causing a further sacrilege. Obaid Allah, the Kufa governor behind the Kerbala massacre, was killed and his head severed as revenge by the rebels supporting the cause of *ahl al-bayt*. Yazid died in 683 CE in the middle of the

---

was prevented by someone, and therefore both Ali and his son were spared. Zain al-Abedin was a title given to Ali later.

[4]The disputed burial places include Kerbala, Damascus, Najf (where Khalifa Ali was buried) and Cairo (apparently reburied there by the Fatimids in a later century).

civil war, which did not subside until the reign of Abd al-Malik (685-705), cousin of Yazid. In between there were two more Khalifas, the second one was Marwan, the corrupt vizier of Uthman, who was succeeded by his son Abd al-Malik. To his credit, Abd al-Malik produced an authorised written-form of the Quran with the latest Arabic script (as mentioned earlier), and also constructed in 691 CE the Dome of Rock Mosque in Jerusalem[5], which was later improved by Salah al-Din Ayyubi (1137? -93), known as Saladin in the West. Abd al-Malik also has a dubious distinction of being the first Khalifa who, following the Byzantine custom, brought in eunuchs to guard the Harem. His wife Atiqah was the great Umayyad queen of style and glamour, but was upstaged a century later by Zubeida[6], the queen of Abbasid Harun al-Rashid (see later). The next Umayyad Khalifa was Walid I (705-715), who in 711 CE extended the Umayyad empire both to Sind in India and to the Pyrenees in Europe. It is this Sind link with India that later provided the Abbasid Khalifas access to ancient Indian science.

After Walid, there were further civil wars and a number of short-lived Khalifas, including Umar bin Abdul Aziz (717-720), who was an exception. He was hailed by the people as the second Umar and as the fifth Khalifa of the Rashidun for his piety and honesty. However, he was murdered by the ruling elite for *not* acting in favour of their interest. The next important Khalifa was Hisham (724-743), who improved the administration and also gave birth to what is now known as the Islamic architecture. The last Khalifa was Marwan II (744-750), who though able, arrived too late to save the dynasty. He was defeated and killed by Abul Abbas, the founder of the Abbasid dynasty.

To give them their due, the Umayyads established a large Arab empire based on sound administration, but they were autocratic, in-

---

[5] As they were hated in Mecca and Medina, the Umayyads projected Jerusalem as the third holy site of Islam. Muawiya not only declared himself Khalifa there, he also undertook rebuilding of the city. Abd al-Malik's construction of the Dome of the Rock Mosque was also partly motivated by this desire to make Jerusalem an alternative holy site. Apparently he also attempted in vain to divert the Muslims from visiting Mecca and Medina, for fear of their being "wrongly" influenced there by the anti-Umayyads. See also footnote 3 in chapter 4.

[6] Valued possessions of Zubeida were a famous heirloom and a ruby-studded dress of Atiqah (rows of huge rubies, both front and back). Zubeida gave the heirloom as a wedding gift to Buran, the queen of Mamun, see the next section.

tolerant of dissent and vindictive towards their opponents. They used the *Baitul Mal* (the public treasury) as their private purse and dispensed it to favour friends. The Abbasids were no better in these aspects, in fact far worse. However, the Umayyads (unlike the Abbasids) encouraged Arab racism, favouring all pure Arabs in high positions, ignoring half-Arabs and non-Arabs, which infuriated the sophisticated Persians. Non-Arab Muslims were required to be adapted into Arab tribes, as honorary tribe members (on grounds that the Prophet and Islam were only for the Arabs). For a while non-Arabs were even prevented from becoming Muslims, because the non-Muslims were a better source of income, as they had to pay more taxes. The Umayyads contrived religion to serve them rather than them serving the religion.

Some translation of Greek works started in their reign, even though they themselves showed no interest in science or philosophy. The exception was Prince Khalid, a son of Yazid, who was personally interested in astronomy and was known as *The Wise*.

2.4 The Abbasids

While the Umayyads ruled for merely ninety years (661-750), the Abbasid Khelafa lasted for over five hundred years (750-1258), the longest reign in Islam except for the Ottomans (chapter 11). During the first hundred years of their reign, the Abbasids created the Islamic civilisation, often referred to as *the Golden Age of Islam*, which saw the rise of science and philosophy, and the development of Islamic law (Sharia). The next four hundred years constituted the period of slow decline, not only of science and philosophy, but also of the dynastic power, the dynasty terminating in 1258 at the hands of the Mongols. We shall cover below both these periods. Needless to point out that the environments that propelled the rise and caused the decline, of science and philosophy under Islam, are rooted in their reigns.

The Golden Age

Abul Abbas, a descendent of Abbas (uncle of the Prophet), led the Abbasids to power with popular support in the name of the House of Ali, which represented *ahl al-bayt* (as mentioned earlier). He exacted revenge on the Umayyads by killing all of them, except a grandson of Hisham called Abd al Rahman who escaped and later established the Umayyad dynasty in Andalusia (Spain). Abul Abbas soon forgot the cause of Ali, took the throne for himself (750-754), brazenly adopting the title *al-Saffah* – the blood-shedder, a title that he lived up to, killing even loyal friends over the slightest disagreements. He kept the theologians on his side by allowing them to help create an Islamic state, although the Khalifa himself and the ruling elite felt free to flout every Islamic code in their personal lives.

While the Umayyads were satisfied by calling themselves Khalifas or vice-regents of the Prophet, the Abbasids went for much loftier titles, emulating the Persian tradition: *Shadow of God on Earth* to imply a divine connection, *Shah-in-shah* (King of Kings) to emphasise the supreme temporal power, and of course *Amirul Mu'mineen* (Commander of the Faithful) to underline the commitment to defend Islam. They also established elaborate Persian court procedures, including one of kissing the ground before approaching the Shadow of God, contrary to Islam. These titles were subsequently grabbed by all major Muslim rulers, including the Ottomans. The Abbasids built a new capital in Baghdad, where they lived in great luxury and opulence[7], beyond the dream of the Umayyads. Furthermore the Abbasids were always surrounded by tiers of court-officials and their underlings who had to be bribed for access to the Khalifa – a Persian practice that ran counter to the very foundation of egalitarian Islam.

Al-Saffah selected his *elder* half-brother Mansur (754-765), son of a Berber slave woman, as the next Khalifa. It perhaps should be noted that the mothers of many Abbasid Khalifas were slave women, often of foreign origin including Greeks. For example, the mother of Hadi and Rashid was a Turkic slave (some say a Berber slave), of Mamun a Persian slave (not the legendary Zubeida), of Wathiq a Greek slave, of Muqtadir also a Greek slave and of Muntasir a

---

[7] Their opulence was legendary. Apparently Zubeida (the queen of Rashid) had a dress of brocade that cost 50,000 gold dinars per square meter (a gold dinar had four gms of gold). The wedding of Mamun to Buran (the daughter of his vizier Hasan) in 824 CE cost 5 million gold dinars for a 17-day festival, in which Hasan showered jewels over the bride and the groom standing on a woven gold mattress, surrounded by the Abbasid princesses. These jewels were meant to be collected by the princesses but they refused until Mamun requested them to honour the host Hasan, at which each princess stepped forward, picked just one jewel and stepped back, leaving most of the jewels still glittering on the mattress [Abb85]. Not even the Indian Mughals of the Tajmahal fame displayed such opulence.

Turkic maid. The mothers of Rashid and Zubeida (his wife) were sisters, both slaves, married to two brothers, one to Mahdi (father of Rashid) and the other to Jafar (father of Zubeida). Therefore, unlike the Umayyads, the Abbasids had much less Arab blood in them.

Returning to Mansur, he was as intolerant and as ruthless as his brother Saffah. As it happens, the Medinans, both Sunni and Shia alike, proclaimed Muhammad al-Nafsal Zakiya, the great-grandson of Hasan the second Imam, as the rightful Khalifa with a view to establishing the House of Ali, as originally promised by the Abbasids. Mansur crushed the pitiful army of Muhammad, killed him and then wreaked vengeance, publicly flogging Imam Malik Ibn Anas and imprisoning until death Imam Abu Hanifa[8] for not opposing Zakiya. He reportedly poisoned the sixth Imam Jafar al-Sadiq in 765. He was also very shrewd and kept the theologians on his side by giving them freedom to create Islamic laws, even though he, his court and his descendents ignored them in their personal lives.

His redeeming features included a profound thirst for knowledge. As discussed in Part II in greater detail, he initiated the Islamic Golden Age, by launching the major project for the translation of Greek and Indian works, and by establishing the medical institute in Baghdad, the new city he built as his capital. However, during this period, Spain was lost to the Umayyad Abd al-Rahman in 756 CE and several other areas also seceded, but he and a number of his successors decided to pour their energies into the creation of an Islamic civilisation rather than into the expansion of the Islamic empire. He was succeeded by his son Mahdi (775-785), an undistinguished ruler, and Mahdi by his two sons, first by Hadi (785-786), and then by Harun al-Rashid (786-809). Rashid was the legendary ruler of *The Arabian Nights* (*The Thousand and One Nights*), who had his grandfather's passion for science and philosophy. He created *Khaznatul Hikmah* (Treasure-House of Wisdom), also called *Baitul Hikmah* (House of Wisdom) in Baghdad, which became the great centre for scholarly activities. Most famous among his advisors was his vizier Jafar Ibn Barmakid, a convert from a Persian family of Buddhist monks. Two women who exerted enormous influence over

---

[8] The imprisonment of Imam Abu Hanifa is disputed. According to his supporters, the refusal of the saintly Imam to accept the earthly post of the Chief Qadi of Baghdad angered Mansur who therefore imprisoned him. Some say he did not die in prison, while some others doubt that he was ever imprisoned.

the Abbasid court were the Khaizuran, the mother of Rashid and a Turkic slave, and Zubeida the wife and double cousin of Rashid. They along with many other women excelled in arts, literature (poetical compositions) and music, the latter despite objections from the orthodox. The most accomplished among these ladies was Ubaydah al-Tunburiah, who became a favourite of the ruling elite and paid little attention to the protests of the orthodox.

In order to avoid a war of succession, Rashid before his death carefully divided the empire between two of his sons: Amin and Mamun. Amin was weak but noble-born (being the son of Zubeida), while Mamun was able but low-born (being son of a Persian slave woman, who died soon after his birth). Furthermore Mamun, who was older by a few months, was raised by Zubeida as a son after his mother's death, and as a result Mamun used to address her as *mother*. In his childhood Mamun outperformed Amin in all princely things, which made him very dear to his father. However, following the plan of Rashid, Amin succeeded to the throne in Baghdad, leaving Mamun, in charge of the outlying areas. Soon disagreement arose and civil war followed in which unfit Amin was defeated and killed by a general of Mamun (apparently without Mamun's authorisation, according Mamun's confession to Zubeida). Mamun had to thwart also a subsequent rebellion of his uncle Ibrahim before he finally succeeded to the throne unchallenged in 814 CE[9]. The removal of Amin ensured greater Persian influence in the court.

Mamun was the greatest patron of science and philosophy, there ever was. While both Mansur and Rashid encouraged the freethinking Mutazilite Movement, Khalifa Mamun actually joined it, and even made it into an official doctrine in 827 CE. The greatest achievements in science and philosophy took place under his proactive stewardship, as narrated in later chapters. He was succeeded

---

[9] The story is quite complex, since at one point Mamun decided to return the throne to the House of Ali, declaring the reluctant 8th Imam Ali Rida (son of Imam Musa Kazim and grandson of Jafar Sadiq) the next Khalifa. He even married one of his daughters to Ali Rida and another daughter (of a different mother) to Rida's son Muhammad (the 9th Imam). At this the Abbasids nominally led by his uncle Ibrahim rose in revolt, which Mamun crushed, but he did not kill his uncle. Imam Ali Rida died of poisoned grapes, allegedly served by Mamun himself. And yet Mamun mourned the death very publicly, buried Rida in a great ceremony next to his beloved father Khalifa Rashid at Mashad (now in Iran) and built there a magnificent mosque as a shrine.

by his half brother Mutasim (833-842), and Mutasim by his son Wathiq (842-847). The Islamic Golden Age began with Mansur and continued until the death of Wathiq. However, during the Mutazila period (827-847), all officials were required to subscribe to the Mutazilite doctrine, including the doctrine of the *created* Quran (see chapter 3). Those who did not conform were removed from their jobs, persecuted, jailed, tortured and in a few cases even killed. This was called *mihna* (inquisition). One person who stood against these Khalifas was Imam Ahmad Ibn Hanbal (see chapter 3).

Wathiq was succeeded by his brother the weak and arch-conservative Mutawakkil (847-861), who in order to please the orthodox, not only banned the Mutazilite doctrine in 850 CE, but also started persecuting them, one victim was Kindi (see chapter 10). The released Imam Ibn Hanbal became the voice of the orthodox and the darling of the uneducated masses, whom he aroused with firebrand speeches to persecute the Mutazilites. Anti-Mutazilite riots followed, leading to the destruction of their properties, livelihoods and even on occasions their lives.

It seems that the seeds of destruction of an empire are usually sown when it is at its height. In the case of the Abbasids, it started right at the beginning of the Golden Age (in the time of Mansur) but it accelerated towards the end of the Golden Age. There is always some rebellion somewhere in any empire. Mansur lost Spain. Mamun's own General who killed Amin, later revolted and established a state in the then Persia. Mutasim felt unsafe in Baghdad and in 836 CE he moved the capital to Samarra (some 60 miles from Baghdad) where it remained for the next eight Khalifas, until it was transferred back to Baghdad in 892 CE. Mutasim started and Wathiq completed the creation of a corps of slaves, mostly Turkic from Transoxania[10], as bodyguards to protect the Khalifa from unreliable Arabs and Persians. Soon the corps became very powerful isolating the Khalifa from his people. Instead of protecting the Khalifa, these bodyguards killed Mutawakkil in 861 CE, their first kill, and then replaced him by Muntasir whom they killed a year later by poisoning. This was the beginning of the bodyguard power.

---

[10] also known as Transoxiana (both meaning the land across the river Oxus) which covered the present-day Kazakstan and part of Uzbekistan.

Decline of the Dynastic Power

The Abbasid history of the next four hundred years is one of general decline in several bursts. The slave bodyguards started making and unmaking Khalifas, killing some, poisoning some, overthrowing some others, and so on, until a Shia general Ahmed Buyid (or Buwayhid) from Iran threw the bodyguards out, and established Buyid rule as the *amir al-amara* (commander of the commanders), but keeping the Khalifa, at that time Mustakfi (944-46), as the nominal head. The Buyids soon blinded Mustakfi and replaced him by Muti' (946-951), thus resuming the act of the making and unmaking Khalifas. To their credit the Buyids, the 12-er Shia (see section 2.6), defended the Sunni Khelafa against the 7-er Fatimids' invasion (see section 2.6) from Egypt in 956. The 12-er Shias hated the 7-er Shias even more than they hated the Sunnis. However, in 1055 the Buyids were replaced by the Seljuk Turks (from Turkistan) as the *amir al-amara*, often calling themselves as the great sultans, under nominal Abbasid Khalifas. Two famous Seljuk sultans were Alp Arslan (1063-72) and his son Malikshah (1072-92). Their distinguished prime minister Nizam al-Mulk Tusi was the patron of Imam Gazzali (chapter 10), but was assassinated by the militant Ismaili Shias led by Hasan Sabbah[11] in 1092.

Salah al-Din, the Great, crushed the Seljuks, returning the Abbasid Khelafa of Nasir (1180-1225) to a flicker of the ancient glory, soon to be extinguished forever in 1258 CE by Halagu Khan, grandson of Mongol Chengis Khan. In the middle of the 13th century the Mongol army destroyed many cities such as Samarqand and Bukhara, killing all their inhabitants, particularly the intellectuals, and burning libraries. In 1258, Halagu Khan, burned and destroyed Baghdad, killed the Khalifa al-Mustasim (by tearing him alive, tied to two horses, or by beating in a sack to avoid spilling the royal blood in another version), massacred most of the inhabitants and destroyed all books and libraries, many by burning. There were too many books to burn, and hence the remaining books were thrown en mass into the river Tigris. It is said that 800,000 people were massa-

---

[11] Hasan Sabbah, also known as *the Old Man of the Mountain* (his hiding place), gave us the term assassin (from hashish smoking) as he wanted to establish an Ismaili Shia state by assassinations. There is a popular story relating Hasan Sabbah, Nizam al-Mulk and Umar Khayyam (chapter 5) as childhood friends who went in their different ways in later life, Hasan's agent killing Nizam.

cred and the water of the Tigris turned blue from the ink of the sunken books and red with human blood. A true act of barbarism.

The Khalifa in Baghdad at that time was holding on to a rump state without any effective military power, full of sound and furies signifying a rotten core of decadence. Some time after 1258, Baibars al-Banduqdari (1258-77), the outstanding Mamluk sultan of Egypt installed a son of a former Abbasid Khalifa Zahir as the new Khalifa in Cairo with the name Mustansir, who in return granted Baibars the official authority to rule over many countries, thus "legitimising" his rule. Although these powerless Abbasid Khalifas had only a ceremonial role, supported as puppets, with a small court and pension, they nevertheless provided useful clouts to the Mamluks. However powerless, the Khalifa was still the recognised Head of the Muslim (Sunni) umma. For instance, the Ottoman sultan Bayezid I (1389-1402) sought and obtained the approval of the Khalifa in Cairo to "legitimise" his title *Sultan of Rum*. This succession of puppet Abbasid Khalifas continued for 250 years until 1517, when the Ottoman sultan Selim I conquered Egypt and moved the then Khalifa Mutawakkil III to Istanbul, where the Khalifa abdicated and then returned to Cairo where he died in 1543 as a commoner (see chapter 11). Baghdad never recovered its former glory; the cultural focus shifted to Cairo, and later to Istanbul.

Summarising, the Abbasids started going downhill from the year 847 when Khalifa Mutawakkil came to the throne. The decline continued with a major loss of power in the year 945 CE when the Buyids took over, but the Abbasids were still reigning as Khalifa although only in name, until the invasion of Halagu Khan in 1258. During this downhill period the empire broke into various pieces, all provincial governors declaring independence, calling themselves sultans, or even great sultans, but never Khalifas, except the Umayyads in Spain and the Fatimids in Egypt (see below). All these sultans recognised the Abbasid Khalifa in Baghdad as the legitimate titular head, and therefore they sometimes included his name as the Khalifa, along with their own names as the sultans, in their coins and the Friday Khutbas (sermons). In the Muslim mind the Khalifa had always this legitimacy as the head of the umma.

Furthermore as the decline continued, the ulama (religious scholars) were gaining power over the rulers and the masses. The Muslims were having a hard time, and therefore return to more orthodoxy was seen as the answer. The rulers needed the support of the ulama to legitimise their rule through the Friday Khutbas; ulama were ready to offer this legitimacy with appropriate fatwas to whoever supported them the most. In Ottoman Turkey the ulama (in fact the Sheikh al-Islam – the highest religious authority) issued fatwas regularly, sometimes sanctioning even the executions of Khalifas or their brothers for political expediencies (see chapter 11). When it suited them, these ulama were capable of issuing fatwas on anything on behalf of the ruling authorities. Religion was used to advance the cause of the ulama. From the official records (e.g. Shahjahan-nama) of the Mughal emperors, it is evident that new rulers used to get very worried about the first Friday Khutba following their accessions to the throne, often after killing their brothers. However, as the ulama gained power, science and philosophy declined. It became the prevailing social view that a good Muslim was not meant to get involved in such ungodly activities as the study of science and philosophy.

2.5 Spain

Abd al Rahman I, grandson of Umayyad Khalifa Hisham, became the new ruler of Spain as its *amir* in 756 CE, after defeating the Abbasid governor Yusuf al-Fahri. As the Umayyad, who escaped the massacre of al-Saffah, Abd al-Rahman promptly denounced the Abbasid Khalifa in Baghdad and refused to recognise him, thus taking Spain outside the main Muslim umma. He made Cordoba his capital where he built the magnificent mosque – with a majestic interior that soothes all eyes, even today after some 1200 years. He was succeeded by his second son Hisham (788-96), which provoked a dynastic war. The unrest continued during the reign of Hisham's son al-Hakam I (796-822), peace returning during the reign of Hakam's son Abd al-Rahman II (822-52). His reign was noted for intellectual revival, particularly in music and art. The succession of his son Muhammad I (852-86) was followed by several uprisings in different parts of the country. While Muhammad was able to suppress these uprisings, they re-appeared during the reigns of his successors al-Mundhir (886-88) and Abdullah (888-912), and were finally crushed by Abd al-Rahman III (912-61), when the country returned to the path of peace and prosperity. He unified the Islamic territories, became the de facto ruler and arbiter of all Spain, declared himself Khalifa and centralised the administration. But the glory did not last for long, his son Hakam II (961-76) was the last great Umayyad ruler of Spain, after which their power started to crumble, the dynasty collapsing in 1031.

Development in science followed the pattern in Baghdad, lagging behind by about 100 years, the role of Mamun (814-833) in Baghdad being played by Hakam II (961-76) in Cordoba. Hakam was very keen to promote science and philosophy in his land – from the time he was the crown prince, when he convinced his father to support his efforts. Hakam sent emissaries to Iraq, Egypt and other countries to collect and copy manuscripts on science and technology. These were brought to Cordoba to create a library which almost rivalled that in Baghdad. However, after his death all books in that library on rational sciences (except arithmetic and medicine) were burnt, destroyed or thrown into the river. Following the fall of the Umayyads, Spain broke into a number of small states, which for a while competed with each other on science and technology. This gave rise to a second period of growth in Spain, paralleling the work of people like Ibn Sina after the decline of the Abbasid power in Baghdad in 945 CE.

The first notable philosopher in Cordoba was Muhammad Ibn Masarra (d 931), who can be compared with Kindi in Baghdad (d 870), though the latter was intellectually far superior. Although the work on science and philosophy began in Muslim Spain perhaps from the 9th century, the major contributions were made mostly between the 11th and 12th century, with only a few exceptions outside this period. Zarqali (astronomy), Ibn Tufayl (medicine), Abu Marwan Ibn Zuhr (medicine), Ibn Rushd (see chapter 10), Ibn Bajja (philosophy), Ibn Hazm (philosophy) were products of the 11/12th century, while Ibn Khaldun (see chapter 10) was an exception. Both Ibn Tufayl and Ibn Zuhr were friends of Ibn Rushd who collaborated with them, and he succeeded Ibn Tufayl as the court physician at Cordoba in 1182 CE. The well-attended and the richly endowed colleges of Cordoba and Granada provided a model and template for those founded later in Oxford and Cambridge in England [Ahm02/p4].

2.6 Shias and Fatimids

Hussain's son Ali and Ali's son Muhammad Baqir, the 4th and 5th Imams, retired from politics in Medina. Jafar al-Sadiq (d 765), the

6th Imam, decided to concentrate on the mystical interpretation of the scripture, instead of fighting for political power, thus separating religion from politics. Before his death in 765, he bypassed his first son Ismail (because of wine-drinking – according to the 12-er Shias) and declared the second son Musa Kazim as his successor, as the 7th Imam. This decision was not accepted by some Shias who regarded Ismail as the legitimate 7th Imam and his son Muhammad, as the 8th. According to these Shias, Ismail predeceased his father, but before his death he was actually nominated Imam by his father, and he (Ismail) in turn duly nominated his son Muhammad as the 8th Imam. This group of Shias is known as the sevener (7-er) Shias, or the Ismailies, which later broke into several further branches, one of which is called Druze, and another is called Nizari. The Nizaris are today led by the Aga Khan (49th Imam) as a non-political movement. In between the Ismailies acted like a secret society, entered into politics, carried out assassinations, and in 969 CE established the powerful Fatimid dynasty in Egypt, rivalling the Abbasids for a while. Hasan Sabbah who caused the murder of Nizam al-Mulk in 1092 (mentioned earlier) was the most notorious secret-society leader who wanted to establish Ismaili rule in Baghdad through assassinations.

The Aga Khan is a Nizari Shia, as he claims to be the descendent of Nizar, who was the eldest son and designated successor of Fatimid Khalifa Mustansir (1036-94), but was overthrown and later violently killed in prison (along with his son), by his younger brother Mustali. Further 7-er branches, including a Mustali branch, followed.

The mainstream Shias are the twelvers (12-er), also called Jafari following Imam Jafar al-Sadiq. Jafar's successor the 7th Imam Musa Kazim died in 799. According to the Shia sources, both Jafar and Musa died of poison, by the order of Mansur and Rashid respectively. The 8th Imam Ali Rida was also apparently poisoned by Mamun (see footnote 9). The 12-er Shias believe that all Imams, except, Ali, Hussain and the last (i.e the 12th) Imam, died of poison by the order of either the Umayyad or the Abbasid Khalifas.

During the time of Khalifa Mutawakkil, when the empire started to crumble, he tried to restrict the freedom of the movement of the Imams. He summoned the 10th Imam al-Hadi and his son (the 11th Imam Hasan al-Askeri) from Medina to Samarra (the new capital since 836, some 60 miles from Baghdad), and placed them under house arrest in 848 CE. Both died in prison, first the father in 868 aged 41, and then the son, the 11th Imam, in 874 aged 27 (probably poisoned by the Khalifa Mutamid's agents). As the 11th Imam lived and died in seclusion, people did not know for sure if he had any sons, but most people believed that he had a son called Abu al-Qasim Muhammad al-Mahdi (*Mahdi* meaning the Guide). He was believed to be Imam Mahdi, the 12th Imam, but in hiding since his birth in 869 CE, for fear of his life from the Khalifa's agents. Since people did not actually see the 12th Imam, a myth soon enveloped his existence and activities. From around 934 CE, he became in popular belief the Hidden Imam, who, as Imam Mahdi, would return to the world shortly before the Day of Last Judgement, fight evil and establish a Golden Age. The world would be corrupt, ruled by bad rulers until the Hidden Imam returned. And therefore there was no point in fighting these corrupt rulers, since none except the Hidden Imam could defeat them. This meant that in the meantime, the twelver ulama should not participate in politics, but concentrate on religious guidance – an attitude that remained unchanged until the early 17th century, when Shah Abbas established the 12-er Shia state in Iran.

Iran: 12-er Shia State

By the end of the 15th century, most Shias were Arabs, most Iranians except those in Qum, were Sunnis. Shah Ismail (1487-1524) was the head of the Safavid sufi order in Azerbaijan, then part of Iran, which at that time covered a very large region. He claimed descent from the seventh Imam and established the Safavid Empire in Iran from 1501, declaring the sevener Shia'ism as the state doctrine [Jac86]. He was succeeded by his son Tahmasp I (1524-76) who strengthened the Shia dominance of the empire and moved the capital from Tabriz to Qasvin. Tahmasp helped the Mughal emperor Humayun to recapture the Indian throne (see section 11.1), though his overture to convert Humayun to Shia'ism did not succeed.

His grandson Shah Abbas I (1588-1629) was the real architect of the Iranian Shia state. He regained territories from the Ottomans and replaced the 7-er Shia'ism by the 12-er Shia'ism. He imported Arab Shia ulama, built religious institutions for them, including many madrasas (religious schools) and successfully persuaded them to participate in the Government, which they had shunned in the past (following the Hidden Imam doctrine). He hated the Sunnis, and forced the population to accept 12-er Shia'ism, thus transforming Sunni Iran into a twelver Shia state, as it is today. He reinforced the doctrine of the Hidden Imam, claiming himself as the deputy of the Hidden Imam, which the Shia clergy did not like. While he succeeded in controlling the clergy, his successors could not. Shah Abbas transferred the capital from Qasvin to Isfahan, encouraged education and science, and initiated a Golden Age which fused Shia'ism with the ancient Sasanid traditions (see section 1.5).

This was the period when the Muslim power reached its zenith with three great Muslim rulers in three different empires, Shah Abbas [Safavid], Akbar the Great [Mughal] and Sulaiman the Magnificent [Ottoman] (see chapter 11 for the last two).

However, in 1736 the Safavids were overthrown by Nadir Shah who then became the King Nadir Shah, abandoned Shia'ism and instead founded a mixed Shia/Sunni school to add to the other four Sunni schools of law. However, the kingdom collapsed after his assassination in 1747 and it suffered destruction under Ahmad Shah Abdali, as did India. After the death of Abdali, Agha Muhammad Khan founded in 1794 the Qajar dynasty, which re-established 12-er Shia'ism as the state doctrine, but without claming any Imam-hood, unlike Shah Abbas. The Qajars gave in to the ulama and created a post of Sheikh al-Islam as the powerful head of the religious affairs (paralleling that in Ottoman Turkey). In 1796, they moved the capital to Tehran and remained in power until 1924, when they were overthrown by one uneducated lowly soldier called Reza Khan, who earlier rose to become a commander of the army and Prime Minister of the last Qajar ruler, Ahmed Shah.

Reluctantly this reform-minded general became Shah (see chapter 12), but in 1939 he was forced by the British to abdicate in favour of his 20-year old son Muhammad Reza, because of his declaration of neutrality in the forthcoming World War. Britain divided the country into two parts. The northern part which included Azerbaijan was given to the Soviet Union, and the southern part (current Iran) was kept under British rule. The son, Reza Shah Pahlavi, ignored the Shia clergy, but was overthrown by them led by Imam Khomeini in 1979. Reza Shah could have got away with his anti-clergy stance, if his administration had not been so autocratic, corrupt and appalling – something akin to Czar Nicholas II of Russia in the early 20th century.

Egypt: Fatimids and Mamluks

In 908 CE, Ismail the 7-er pretender became the first Fatimid Khalifa in North Africa under the name Obaidullah. The Fatimids took their name from Fatima, the daughter of the Prophet, wife of Ali and the mother of both Imam Hasan and Hussain. The fourth Fatimid Khalifa Mu'izz conquered Egypt in 969 CE and built the new city of Cairo as his capital. The dynasty gradually extended its empire into the whole of North Africa, Sicily, Western Arabia, Syria and Palestine, reaching its greatest extent during the reign of the Fatimid Khalifa Mustansir (1036-1094). In 1056, it even captured Baghdad and proclaimed the restoration of the House of Ali for the rightful Khalifas, in place of the "charlatan" Abbasids. But the contraction of the empire and its decline started from the following year, when they were driven out of Baghdad by the Buyid 12-er Shia protector of the Sunni Abbasid Khelafa. After the death of Mustansir, his eldest son and designated heir, Nizar was overthrown by the younger son (mentioned earlier). The Fatimids' powers dwindled fast and in 1171 CE the Ayyubi dynasty was established in Egypt by Salah al-Din the Great, which recognised the Sunni Abbasid Khelafa in Baghdad – titular though the Khelafa was from 945 CE.

The Fatimids established al-Azhar University at Cairo, the great Islamic seat of learning, and under their rule Egypt prospered as never before. But they ruled as autocrats, claiming divine rights. They ignored ulama, favoured Christian and Jewish advisors and detested the Sunni Abbasids. After a while they became oppressive and the majority Sunni revolted against their rule, but their rule continued until they were removed by Salah al-Din. In 1250, the Mamluks took over Egypt. Mamluks were a crack military corps composed of Circassian slaves captured as boys and converted to Islam, similar to the Janissaries of the Ottomans (chapter 11). They ruled Egypt until 1517 when Egypt achieved great heights in science and technology.

The Mamluk rule was crushed in 1517 by the Ottoman Khalifa Selim I in league with Khair Bey, the Mamluk governor of Aleppo, who was subsequently made the Ottoman governor of Egypt. Thus the Mamluks did not disappear, in fact they continued to rule Egypt right up to the end of the 18th century, often in open rebellion against the Ottomans. This rule of the second set of Mamluks was disastrous for Egypt. The fact is that the Beys (commanders) who controlled these Mamluks were often involved in internecine strife, caring very little for Egypt or the Egyptians. They were oppressive – they plundered Egyptian homes, confiscated their property and crippled them with taxes. Under them Egypt suffered greatly, lost all its earlier gains in science and technology, and became a closed backward province of the Ottomans. They were removed from power by Napoleon, but finally eliminated by Muhammad Ali (see chapter 12), whose dynasty ruled Egypt until 1952, when its last king Faruq was overthrown by Gamal Abd al-Nasser.

References and Sources

[Ali68] Syed Ameer Ali: The Legacy of Islam, first published in 1931, reprinted by Oxford University Press in 1968.

[Abb85] Nabia Abbott: Two Queens of Baghdad, Saqi Books, London 1985. The book describes the Abbasid story from Mansur to Mamun and their queens.

[Abb98] Nabia Abbott: Aishah – the Beloved of Muhammad, Saqi Books, London, Republished 1998. These two books of Nabia Abbott are very illuminating on the customs, practices, culture, affluence and cruelties of the period. She is also a hadith expert.

[Ahm02] Akbar Ahmad: Discovering Islam, Routledge, 2002.

[Arm01] Karen Armstrong: The Battle for God, HarperCollins, 2001.

[Arm91] Karen Armstrong: Muhammad – A biography of the Prophet, Gollancz, 1991. The book is based on the earliest sources.

[Arm92] Karen Armstrong: Holy War, Papermac, 1992.

[Bri 71] Encyclopaedia of Islam, ed E. J. Brill, (Leiden 1971), Vol 3.

[Esp98] John L. Esposito: Islam – The Straight Path, Oxford University Press, 1998.

[Epi99] John Esposito (ed): Oxford History of Islam, 1999.

[Gru70] G. E. Von Grunebaum (ed), Translated by Katherine Watson: Classical Islam, George Allen & Unwin, London 1970.

[Gui87] Alfred Guillaume: Islam, Penguin, reprinted 1987, pp106-10.

[Gui95] Alfred Guillaume: The Life of Muhammad – A Translation of *Sirat Rasul Allah* of Ibn Hisham (*This is the book on the life of the Prophet from the earliest sources*), Oxford University Press (Karachi), Tenth Impression 1995.

[Jac86] P. Jackson and L. Lockhart: Cambridge History of Iran, Vol 6, Cambridge University Press, 1986.

[Ham62] Hamilton Gibb: Studies on the Civilisation of Islam, Princeton University Press, 1962, p142/3.

[Holt70] P. M. Holt et al (ed): Cambridge History of Islam, Vol 1, Vol 2a, Vol 2B, Cambridge University Press, 1970.

[Hou02] Albert Hourani: A History of the Arab People – A New Edition, updated by M. Ruthven, Faber & Faber, 2002.

[Lew97] B. Lewis: The Middle East, Phoenix Giant, 1997.

[Lin91] Martin Lings: Muhammad – his Life Based on the Earliest Sources, Islamic Texts Society, 1991. [Lings is a well-known Muslim sufi, and received awards from many Muslim countries for this book].

[Rog04] B. Rogerson: Prophet Muhammad – A Biography, Abacus, 2004.

[Rog06] B. Rogerson: The Heirs of the Prophet Muhammad, Abacus, 2006.

[Rut91] Malise Ruthven: Islam in the World, Penguin 1991.

[Tab99] Abu Jafar Al-Tabari [d 923]: *Ta'rikh al-Rasul wa al-Muluk* (History of the Messenger and Kings), translated in 39 vols, SUNNY Series in Near Eastern Studies, Albany State University of New York Press, 1985-99.

[Wal06] T. Wallace-Murphy: What Islam Did for Us, Watkins, London 2006.

[Wat96] W. M. Watt: A History of Islamic Spain, Edinburgh University Press, 1996.

[Zak89] R. Zakaria, The Struggle Within Islam, Penguin, 1989.

CHAPTER 3

THE MUTAZILA MOVEMENT

The ideas advocated by the Mutazilites can be traced back to the earlier battle during the Umayyad period on *Freewill* fought by a sect called Qadarites (*qadar* – destiny), who used Greek logic and rationalism to propagate their cause. How a philosophical concept like freewill became the enemy of the state is an interesting commentary on the nature of the Umayyad state. The stimulus to develop the concept of freewill arose from the need to portray God as just. If everything is pre-destined, including the commission of a sin by someone, then how could a just God punish that person? Thus rationalism and Greek logic demanded the existence of freewill, as advanced by the Qadarites.

There was also a counter movement called Jabria, which believed in pre-destination. According to them the brutal beheading of Imam Hussain at Kerbala on behalf of Yazid (the second Umayyad Khalifa) was also pre-ordained by God, and therefore Yazid was not responsible. This implied that the tyrannical regime of the Umayyads is also pre-ordained, and hence should not be opposed. The freewillers took the opposite stance and hence became the enemy of the Umayyad dynasty. Both groups, Qadarites and the Jabriates, used Quranic verses to support their respective positions.

Angered by the Qadarites, Umayyad Khalifa Abd al-Malik beheaded its leader Ma'bad al Juhani in 699 CE at Basra [Hoo91]. This was followed by further executions, hangings and torture of other Qadarites, and also by violent clashes between them and the Jabriates in Basra. This resulted in the demise of the Qadarite school of thought, but from its ashes arose the Mutazilites (the dissenters) with freewill as part of their core beliefs.

The Quran urges Muslims to use their intellect and reason to understand God and His creation. This inspired Muslim intellectuals in the early 8th century to apply their knowledge of Greek logic and philosophy to build a systematic theology based on a rational foundation. The man who is usually credited as the founder of the Mutazila school in the early 8th century was Wasil Ibn Ata (699-748), a sufi and former disciple of Imam Hasan al-Basri (641-728 CE) of Basra. Hasan, an ascetic originally from Medina and later revered as the father of sufism, was concerned with the concept of divine justice, freewill, good and evil. Like the Qadarites, Hasan believed in freewill, but accepted the Umayyads as the legitimate rulers [Arm01], while reserving the right to criticise them when they were wrong. After a disagreement with Hasan on the state of a sinful believer in Islam, Wasil Ibn Ata broke away to establish the school of "Mutazila" (meaning dissenters) with the help of Amr Ibn Ubayd (699-761), in order to reconcile faith with reason. The Mutazilites, as they insisted on a primary role for reason in religious dogma, were also called *ahl al-tauhid*, the upholders of the doctrine of the strict oneness of God. This new school soon attracted both Sunni and Shia followers alike, but mostly from non-Arab stock, particularly Persians.

3.1 The Basic Mutazilite Philosophy

Nearly a century later another adherent Abu Al-Hudayel (751-841) provided a clear formulation of the five Mutazila principles: (1) Unity of God, (2) Divine justice, (3) the Promise and threat, (4) the Intermediate position and (5) Commanding of good and forbidding of evil [Hou85, Lea85]. The first two principles were of major importance, the third is a corollary of the second, the fourth is about the intermediate state between a believer and a non-believer, and the fifth is about the obligation of all Muslims to intervene in the affairs of the state, based on some Quranic verses[1]. This fifth principle required the Mutazilites, who were opposed to court luxury [Arm93], to be politically very active.

In addition they also constructed a model for cosmology in which all bodies in the universe are made of atoms, with groups of atoms (equivalent to molecules in today's scientific terminology) giving special properties, such as colour or taste. Some of the Mu-

---

[1] such as ayah [9:71], which urges all believing men and women to "enjoin the doing of what is right and forbid the doing of what is wrong".

tazilites viewed atoms as indivisible mathematical points. These ideas were used to define the difference between the one and the same physical body of a human being at different times, and to explain the application of divine punishment on these different time-versions of a physical human body. However, for us in this book, the first two of their five principles are more interesting, and hence are explained in more detail below.

The Quran asserts very strongly the doctrine of the unity of God (Quran [112 (sura Ekhlas)]). It also speaks of God's *hands* (Quran [38:75]), *eyes* (Quran [54:14]), *face* (Quran [55:27]), and of *He himself sitting on a throne* (Quran [20:5]), thus apparently implying a physical body of God. The Mutazilites rejected this idea of a body of God, arguing that such a physical body can belong to only objects in the universe, but not to God, who according to them is merely numerically one – an abstract entity. They interpreted Quranic anthropomorphism to be metaphors for quality, thus *hands* meaning His blessings, *eyes* His knowledge, *face* His essence, and so on. There is a verse in the Quran [3:07] which says: ".... and others [i.e other verses] that are allegorical ... ", implying that some of statements in the Quran are allegorical. This can be used to justify the Mutazilite position.

There are some adjectives of God in the Quran, that were classified as the *attributes of essence*. The Quran describes God as *living, knowing, powerful,* and *eternal*. According to the Mutazilites these are the attributes of perfection, viewed from the different aspects of the same perfect one God, and hence these are *the attributes of essence*. The Mutazilites viewed God as pure essence, impossible for a human being to comprehend as stated in the Quran [2:02/03] (given below). This view also ran counter to the orthodoxy, that took God's attributes in the Quran literally, and accused the Mutazilites for denuding God of content, and thus making it difficult for the Muslims to comprehend and worship Him. On the other hand the Quran says:

"[the Quran] is the guidance for all the God-conscious who believe [in the existence] of that which is beyond the reach of human perception ...." . Quran [2:02/03].

The Mutazilites identified a third kind of adjective which relates to God's actions, as the temporal attributes of God, but not of essence. These attributes are created when God acts, but not before. One such attribute is God's speech. For example, if God asks one of His creations to obey a commandment, then it does not make sense to have that commandment existing before that creation is created. Thus the Quran, the Word of God, is temporal, *not* eternal. This is the Mutazilite doctrine of the *createdness* of the Quran (*Khalq-i-Quran*), which says that the Quran is the created rather than uncreated Word of God. This doctrine they argued did not contradict the ayah, Quran [85:22], which says that it (the Quran) was "*inscribed [or preserved] in a well-guarded tablet*". According to them, this means that the Quran is imperishable, that is, free from corruption.

However, if the Quran is uncreated (the current belief of most Muslims), then it would have the same attributes as those of God – it would be eternal and perfect. In that case, the Mutazilites argued, it would be equivalent to God, thus violating the unity of God in Islam. Since the Quran had a historical context (for example, the speeches of Moses in the Quran), this was a further argument in favour of createdness, according to the Mutazilites. This idea of createdness was perhaps the most important doctrine to have come out of the Mutazilite school, and it has implications for the Muslim community today, if it wishes to move forward in the modern world and in the pursuit of science and philosophy. We shall therefore examine it a little further shortly.

Divine Justice and Freewill

God rewards the virtuous and punishes sinners. God is Omnipotent and also Just. Without *freewill* to act, a human being cannot be responsible for his actions. If God predestines a human being to be evil, then He cannot punish the evil-doer, if He is to be just, merciful and compassionate. Now consider the question: could God send a virtuous man to hell? Logic of justice requires the answer to be *no*, but God's omnipotence demands the answer to be *yes*, since omnipotence implies that God's power to send someone to hell is not limited even by the need for justice. Therefore, justice necessitates not only freewill, but also God's omnipotence to become limited by His own nature. Thus in the Mutazilite doctrine, God is the maker of laws, but not an executor of arbitrary unlimited power. Obviously the orthodoxy could not accept such a conclusion. In their view, if man has freewill to act, then he is the creator of his action, which is blasphemous, since only God is the creator of all things.

Much later a non-Mutazilite solution was put forward by Maturidi of Samarqand (see also below) in which God is the Creator of both good and evil acts, but man has freewill to choose (God *creates*, man *acquires*). This was not acceptable to either side. In many places in the Quran, God urges believers to act – to pray, to keep fast, to do good deeds and so on. God warns believers of the consequences of doing evil. How could these assertions of God make sense if man does not have freewill to act – argued the Mutazilites. While there are ayahs (i.e. verses) in the Quran which say that nothing can happen without the permission of God, there are also other ayahs which seem to support freewill:

"Had it been Our will [that men should not be able to discern between right and wrong], We could have deprived them of their sight so that they strayed for ever from the [right] way …..". Quran [36:66].

"Had it been Our will [that they should not be free to choose between right and wrong], We could surely have given them a different nature ….". Quran [36:67].

"Verily God does not change the state of a people unless they change the state of their inner selves". Quran [13:11].

These verses of the Quran can be interpreted as supporting the Mutazilite concept of freewill. With historical hindsight, we can say that the philosophical issues that bothered the Mutazilites greatly, were the meaning of reward and punishment, the relationship of God to His Word (the Quran) and God's nature and attributes. They were devout Muslims, and wanted to understand their religion intellectually, following a saying of the Prophet that intellect is the greatest gift of God to man. There are also many verses in the Quran exhorting man to reflect and understand. One such verse examined in the next chapter is:

"… He sent down to you the Book and Wisdom for your Instruction. …".   Quran [02:231].

3.2 Doctrine of Khalq-i-Quran

The Mutazilites did not believe hadiths to provide the only possible interpretation of the Quran, in fact they went even further and refused to use hadiths for the interpretation of the Quran at all, because, they argued, hadiths are essentially unreliable and full of contradictions. Some examples of these contradictions as cited by them are given in [Gui87/p106-110]. Instead of hadiths, they decided to depend on the other instrument, namely, human intellect, which they perceived to be the ability for rational analysis through reason, based on Greek logic. They celebrated this ability as God's gift, following the verse [02:231 cited above. For more details on hadith issues see chapters 4 and 13.

The Mutazilites quoted from the Quran to support their doctrine, as did those who opposed them. The principle of *Khalq-i-Quran*, as stated above, was most powerful in its implication for the acquisition of knowledge and in the development of both laws and science under Islam. If any verse in the Quran has historical context, then the Quran will have historical elements in it, and therefore it cannot be eternal. All readers of the Quran will be able to find numerous verses with historical context, some of which are cited below:

1. Abrogation:
"Any verse that We annul or consign to oblivion, We replace it with a better or similar one". Quran [2:106].

This implies that the Quran has been changed (and hence created) by God at some point in time and therefore it cannot be eternal.

2. Wine: there are three relevant verses
"And from the fruit of date-palms and the vine you get out strong drink as well as wholesome food". Quran [16:67].

"They will ask you about wine and gambling. Say in both there is great evil as well as some benefit for man; but the evil they cause is greater than the benefit they bring". Quran [2:219].

Some ten years after the verse [16:67] above, came the verse [5:90]:
"Oh you who believe! Intoxicants, gambling, idolatrous practices and divination of the future are an abomination – of Satan's doing: shun it so that you might prosper".

Some people argue that the term *shun* does not have the force of *haram* (i.e. forbidden). It is also claimed, the rule on wine was tightened gradually for the society to get used to it. In that case these verses have a historical context, implying that God changed the rule on wine over ten historical years, and hence the content of the Quran cannot be eternally true. This can also be used to justify new interpretations of the Quran over time, and the introduction of new rules as society changes.

3. Creation of the Quran

> "Now this Quran could not possibly have been devised by anyone save God" Quran [10:37].
>
> "Behold we have sent it down as a discourse in Arabic, so that you can encompass it with your reason" Quran [12:2].
>
> "We have made it in Arabic" Quran [43:3].
>
> "... We have sent down the Message ..." Quran [15:9].

So it was created by God, not uncreated, unlike say His *power*. Also God could have sent it in another language, and therefore another Quran was possible. It was made in Arabic, implying creation. Also a message is something that is created. Finally, it does not make sense to believe it was created in Arabic before the Arabic language was created, and hence Quran must be temporal. There are also many other ayahs which one can cite in favour of createdness. It seems Imam Abu Hanifa who died in 767 CE held a somewhat ambiguous position on createdness [Zak91]. In some modern Muslim rethinking, the controversy of createdness is neatly bypassed by considering a new form of contextual interpretation of the divine Word presented within the constraints of a human language (see chapters 13 and 14).

3.3 Asharite Opposition

Later another school called Asharites was established by Abu al-Hasan Ibn Ismail al-Ashari (878-941), a former Mutazilite who subsequently went to the other extreme after a dream in which the Prophet had asked him to study the hadiths (Mutazilites did not trust hadiths). He then had another dream in which the Prophet said: "I did not tell you to give up rational arguments to support true hadiths", which led him to use rationalistic techniques, detested by the traditionalists, such as Imam Ahmad Ibn Hanbal. The Asharites went on to develop a compromise doctrine between the Mutazilites and the traditionalists, based on reason and logic, in which God is beyond our understanding. However they were fundamentally opposed, as were the traditionalists, to science. Their doctrine later led to the philosophy of *ilm-al kalam* (knowledge of kalam, or just *kalam*[2]), literally knowledge of discourse, in which the Quran is uncreated and God is beyond logic, reason and human perception. Therefore, God could indeed be seated physically on a throne, with all its glory (Quran [20:5]), even though we may not be able to conceive of a pure spirit sitting – we must *not* try to question or rationalise this as an allegory.

Ashari wanted to explore metaphysical issues, relating to the nature of God through *kalam*, even though he finally concluded the reality of God to be indescribable, incomprehensible and inaccessible to the human mind. The first major theologian of the Asharite school was Abu Bakr al-Baqillani (d 1013), but the most prominent follower was Imam Gazzali (d 1111, see chapter 10). The doctrine was attacked by later traditionalists for its use of logic. Profoundly opposed to science, Asharism often employed over-literal interpretation of essentially illusive religious issues, without any evidence-based reasoning.

While the Asharites did not believe in freewill, al-Maturidi (d 944), a great Hanafi theologian from Samarqand, conceded to a doctrine of limited freewill (the current Sunni belief), as a moral basis for reward and punishment, declaring God to be bound by an absolute ethic. Thus a limited form of freewill was restored.

3.4 Fall of the Mutazilites

Khalifa Mansur encouraged the Mutazilite doctrine, as did Khalifa Rashid, but the latter's son Mamun elevated it in 827 CE to a state doctrine, which it remained throughout the time of Khalifa Mutasim (833-842) and Wathiq (842-847) until 850 CE when it was aban-

---

[2] The term kalam is often interpreted as just theology, and hence the Mutazilite doctrine is also described sometimes as the Mutazilite kalam.

doned by Khalifa Mutawakkil (847-861). However, it failed to penetrate the depth of the uneducated Muslim masses, who found such abstract thought incomprehensible and unacceptable. Like all philosophers of the time, the Mutazilites did not care much for the masses; they were proud that their doctrine was meant only for the intellectuals who could understand higher things beyond the grasp of the masses.

During the time of Khalifa Mamun, the doctrine became attractive to the court and to educated people, without much intellectual challenge. It was studied in madrasas (religious schools) and mosques, and it spread to all parts of the Islamic empire, even to Persia and later to Andalusia (Spain). Rival schools grew in Baghdad and Basra on the alternative aspects of the doctrine. But there were also excesses and oppression by the rulers.

Khalifa Mamun required all officials to express belief in *Khalq-i-Quran*, and he launched an inquisition (*mihna*) against those official and important religious leaders who refused to accept the doctrine. Many were tortured and imprisoned, some even died. In the absence of any serous intellectual challenge, the opposition was negligible to start with, but it galvanised later, led by the orthodox Imams. One formidable opponent was Imam Ibn Hanbal (780-855), a literalist who refused to recant his orthodoxy even under torture to the point of death. When he was released by Mutawakkil in 850 CE after 16 years in prison, he instantly became "a hero of the masses, a saint, the scourge of rationalists" and "the saviour of orthodoxy and freedom of conscience and faith in Islam".

Imam Hanbal was a traditionalist with an extreme position, and he said: "every discussion about a thing that the Prophet did not discuss is an error". He claimed God to be beyond all logic and conceptual analysis, even though the Quran repeatedly urges the Muslims to apply intelligence and understanding. In this sense his position lacked the intellectual rigour needed by the educated, but it had enough raw power to rouse the less educated masses. Syed Ameer Ali (19th century Indian reformer – see chapter 12) accused Imam Hanbal as the one who manipulated the ill-educated backward Muslim masses, describing him as a:

"red hot puritan, breathing eternal perdition to all those who differed with him. ... He denounced learning and science and declared a holy war against rationalism ...". [Ali67/p439].

But the Muslim masses loved him and his "fire and brimstone" against rationalism – the streets of Baghdad became frequent scenes of riots and bloodshed.

Khalifa Mutawakkil, who abandoned the Mutazilite doctrine in 850 CE, also razed the tomb of Imam Hussain in Kerbala to the ground as an anti-Shia act, and was described by Syed Ameer Ali as the "cruel drunken sot, almost crazy at times " [Ali67/ p439]. The persecution of the Mutazilites began soon after their loss of royal protection. The Mutazilites and Shias were declared heretics, removed from jobs, tortured and sometimes summarily executed. Scholars and scientists, who were sympathetic to the Mutazilites, left Baghdad for their own physical safety. However, Mutazilite scholarship continued in Basra, which culminated in the later work of Abdul Jabbar (935?-1025 CE), after which time there was a diminishing number of Mutazilite scholars. But the intellectual deathblow was delivered subsequently by Imam Gazzali (see chapters 9 and 10). The doctrine did however have some influence up to the 13th century. The Basran school lasted longest, disappearing after the Mongol invasion of 1258 CE. Even today it still continues to influence the Shias in Iran and the Zaydis in Yemen. Needless to say other Muslims, particularly the Sunnis, regard the Mutazilite doctrine as a heretical one.

3.5 Concluding Remarks

The Mutazilite doctrine of *Khalq-i-Quran* was particularly unacceptable to the Muslim orthodox. This doctrine implied that like any creation of God, the Quran can also be imperfect (in the sense that it might not have included all the laws ever needed for all human societies for all times). In that case, the Khalifa, as the supreme Muslim leader, could create laws, independent of the ulama (religious scholars). On the other hand if the doctrine of the uncreated Quran is held to be true, then the Quran has all the perfect answers for all times (including of course all laws). In that case, it would be up to the ulama to interpret the Quran appropriately and determine all laws. This was a battle comparable to that of the reformation during the Christian Church in Europe in the middle ages. The Pope lost his authority to the various Protestant movements and then to state powers, which led to the opening up of societies and the development of science and technology in Europe. In Islam that battle was lost to the religious orthodoxy by the autocratic state in 850 CE.

After the fall of the Mutazilites the pro-hadith ulama, inspired by earlier works of Imam Shafi'i, developed an Islamic theology and Sharia law based on the strict orthodox interpretation of the Quran and hadiths, proudly denouncing rationalism (chapter 4). The result, we would argue, is the current Muslim backwardness in science. The autocratic state under the influence of the Mutazilite doctrine created a Golden Age in Islam through developments in arts, literature, philosophy and science. Could that state have evolved, if it had triumphed over the orthodox, into a state that had dynamism, liberalism and other structures needed for modern intellectual and scientific pursuits? This is a debatable question to be examined in Part III of this book.

References and Sources

[Ali67] Syed Ameer Ali: The Spirit of Islam, first published in 1922, reprinted by Methuen & Co. in 1967.

[Arm93] Karen Armstrong: A History of God, Mandarin paperback, 1993. [A good short section on the Mutazilite movement].

[Arm01] Karen Armstrong: Islam – A Short History, Phoenix paperback, 2001.

[Gui87] Alfred Guillaume, Islam, Penguin, preprinted 1987.

[Hoo91] Pervez Hoodbhoy: Science and Islam, Zed Books, London, 1991.

[Hou85] G. F. Hourani: Reason and Tradition in Islamic Ethics, Cambridge University Press, 1985.

[Lea85] Oliver Leaman: Introduction to Medieval Islamic Philosophy, Cambridge University Press, 1985 [provides a critique of the Islamic rational philosophy].

[Von 70] GE Von Grunebaum, The Horizon of Islam ... in Classical Islam, translated by Katherine Watson, published by Allen and Unwin, London, 1970 [It has some general comments on the Mutazilites].

[Rut91]: Malise Ruthven: Islam in the World, Penguin 1991. [A deep book on the philosophical ideas of Islam].

CHAPTER 4

HADITHS AND SHARIA

After the death of the Prophet, the Rashidun used the Quran and the examples of the Prophet as guidance to handle ever increasing new situations. Where necessary they also deviated from the Prophet's examples for greater good, recognising the special context of the examples and the changed context of new situations. As the time passed with new generations of Muslims, the ad hoc legal decisions gave rise to locally based laws all over the expanded Islamic empire during the Umayyad period (661-750 CE). However, this created an anomalous situation in which different judges (qadis) particularly from different localities (such as Medina or Kufa) could give conflicting judgements on the same case, all equally valid[1]. Since all Muslims formed part of the same umma, there was no concept of area jurisdiction of courts – anyone could select any judge from any court anywhere in the empire. Not only was there no notion of area jurisdiction of courts, there was also no concept of moderation by superior courts or of precedents[2]. With the enlargement of the empire, these conflicts and variability became too serious to ignore. The solution was seen in the formulation of a more universal law, necessarily based on the Quran.

---

[1] Arwa (known as Umm Musa), wife of Abbasid Khalifa Mansur, had in a written stipulation in her marriage contract that Mansur could not take any other wife, not even a concubine, during her life time. When Mansur wanted to wriggle out of this stipulation, Umm Musa brought a qadi from Egypt to give a verdict in her favour, which he did, forcing Mansur to remain monogamous until her death. Apparently Umm Musa's bribe was bigger, but nevertheless this is another example of the absence of area jurisdiction in Islamic law. It is said that after her death, at the tenth year of his reign, Mansur was offered 100 virgins by his grateful subjects! [Abb85]
[2] These issues remain problematic in Sharia law even today.

While everyone accepted the Quran as the basis of such a law, there were heated debates on the method of its interpretation, a battle that was eventually won by the ulama. They believed that only sunna (the apostolic tradition) and hadiths (sayings of the Prophet) provided the valid (in fact the divine) interpretation of the Quran. Thus sunna, hadiths and law are closely related in Islam. However, observe that the resultant law, called *Sharia* (sometimes Sharia law), covers not only civil and criminal laws, but also laws on religious matters, such as, religious rites, rules, duties and obligations. Since Sharia law has shaped the everyday life of Muslims and their attitudes to science and knowledge, we shall review it here. Note that some of the ancient Muslim scholars mentioned below will be visited again more fully in later sections, depending on where their major contributions have been discussed.

4.1 Hadiths

While a hadith of the Prophet is his saying, there can be hadiths of other people as well. Of particular interest to us are the hadiths of his important Companions related to what the Prophet said or did, and also concerning what he approved or disapproved. A hadith, which was originally meant to be an oral report, has two parts: *isnad* (transmission-chain) and *matn* (text). The *isnad* describes the successive transmitters of the *matn* linking the current transmitter to the source (which in the case of the Prophet is one of his Companions). The earlier hadiths did not have *isnads*. These were added later, starting from the turn of the 1st century AH (7th/8th century CE). A sunna of the Prophet will normally be supported by a hadith of the Prophet or of a Companion.

In this book, a hadith will mean a hadith of the Prophet, unless otherwise qualified. The *isnads* of these hadiths go only up to the Companions who are believed to be telling the truth. However, while the Quran had been most carefully preserved from the beginning, hadiths had not. In fact the Rashidun forbade the collection of hadiths (see chapter 13). Despite this injunction, the need for the documentary evidence for hadiths became paramount when the decision was made to base Sharia on sunnas and hadiths in the third century of Islam (9th century CE), but by then hadiths had become riddled with forgeries. An *isnad* verification technique was applied to distinguish the reliable hadiths from the unreliable ones, eventually leading to the compilation of the oral hadiths into books on collections.

Hadith Fabrication

It seems the Prophet was aware of hadith fabrication. According to a hadith of Ali:

> "The Prophet himself had had to issue the warning, 'Do not father lies on me, for he who does will enter hellfire' ".

There are several other Companions (for examples, Anas, Abu Huraira) testifying to similar warnings by the Prophet. Abdullah Ibn Abbas (cousin of Ali and ancestor of the Abbasids) stated: "After Ali's death, they invented all manner of things which wrung from one of his associates. 'Damn these people'! They perverted so much religious information' ...". A deliberate attempt to manufacture fake hadiths for political purposes started during the reign of Muawiya, the first Umayyad Khalifa, who asked his followers to produce hadiths in support of his dynasty and administration. His successors pursued this practice[3]. The opponents responded by generating counter-hadiths.

Prophetic endorsement was considered very important for other areas as well. For example, people proposing a scientific theory, would back it up with hadiths, such as, the sky was made of white marbles or from a jade stone, or that the sizes of the sun and the moon were the same, and so on. There were also many curious (false) translations of the Quran supporting specific points of view. Among the worst offenders were the popular preachers and the pious promoters who exaggerated the contents of hadiths and sometimes fabricated them to enforce religious piety. The following hadith was apparently manufactured after the fall of the Mutazillites:

---

[3] It is claimed that the famous hadith which says that a prayer in the "farthest mosque" (implying al-Aqsa in Jerusalem) is like 1000 prayers elsewhere (except in Mecca and Medina) was manufactured for Khalifa Abd al-Malik, who wanted to encourage Muslims to visit his new mosque (Dome of the Rock) in Jerusalem, instead of going for pilgrimage to the rebel territories of Mecca and Medina, where they could become contaminated with anti-Umayyad ideas. During the civil war, he attacked Mecca flattening even the Kaaba.

"He who comments on the holy texts on the basis of reasons and arrives at the correct interpretation, arrives at the wrong interpretation".
- Al-Tabari (d 923 CE) – Tafsir of the Quran, Vol 1, pp77-79.

This was designed to instil in the minds of the believers the traditional interpretation of the Quran, as against rational interpretation advocated by the Mutazilites. According to Yahya bin Sa'id (see below), many persons supporting a particular doctrine created hadiths complete with *isnads* to promote their cause. Some of them later recanted their fabrications when they changed their doctrines.

Imam Ibn Hanbal accepted any hadiths that supported his views, Imam Shafi'i accepted any hadith that seemed reasonable to him, while the Hanafi's were, at least initially, more strict. Given so much uncertainty about the correctness of hadiths, techniques were needed to separate the sound hadiths from the weak ones. In the 9th century, a number of scholars, such as Yahya bin Sa'id (737-813), Abdul Rahman bin Mahdi (752-808), and specially Abdullah al-Madini (777-848), formulated comprehensive techniques for the verification of *isnads*, and found many inconsistent (and hence presumably fabricated) hadiths. Their general idea, as contained in their technique outlined below, was adapted by others but without rigour.

Hadith Validation Techniques and Practice

Even though initially wholly oral, some hadiths were later held in written form, but often without recording the *isnads* carefully. Thus, while the written form avoided the memory-loss on the *matns*, it did not secure reliability of the *isnads*. Furthermore it also suffered from textual alterations due to the evolution of the Arabic scripts between the 7th and the 9th century CE. A teacher would normally recite (or cause to be recited) his written notes of hadiths to his group of pupils, who (when they became teachers) would in turn recite these to their pupils. Sometimes, a teacher would hand over the notebook to his pupil. A notebook might have contained hadiths from many different predecessors, generally without proper *isnads*, thus making it hard for the later experts to be sure of the correct *isnads*.

Observe that in principle a hadith (on a topic) may have many versions (or reports), each version having a different *isnad*. Therefore the *isnad* and the *matn* (text) from different versions of a hadith can be compared for validation. As regards the grading of *isnads*, the following terms were proposed:

adil: the transmitter is most pious Muslim

thiqa: the transmitter is most reliable

thabit: the transmitter is most accurate

hafiz: the transmitter has good memory

If a hadith (on a topic) is transmitted by several persons (that is, several versions), then it is reliable, and if the text is also identical, then most accurate. A hadith is sound (*sahih*), if it satisfies all the four *isnad* criteria (*adil, thiqa, thabit* and *hafiz*). Next to *sahih* is *hasan* (fair), which is acceptable in the absence of a superior report. If some versions of a hadith are *sahih*, but other versions are *hasan*, then this hadith can become *hasan sahih*. A hadith is *da'if* (weak), if the transmitter is known to have made many errors. It is always assumed that the errors were made due to either carelessness or faulty memory, but never deliberately, even though such an assumption was often unsound. The Mutazilites were scathing about this assumption. A weak hadith is sometimes accepted in the absence of any better ones. The hadiths of the fourth category are those for which there was clear evidence of fabrication, and hence were not acceptable. However, these criteria have been applied only nominally (without rigour) in different hadith compilations, each collector defining the terms subjectively as he wished, as we shall see soon.

Isnad verification was essentially a statistical validation. All versions of a hadith were collected, their *isnads* compared and examined and then a conclusion is reached on both the *isnad* and the *matn*. In general, *isnads* were accepted in good faith, the checks made enhanced the probability of correctness, but it did not guarantee absolute correctness. Since, hadiths were normally transmitted by teachers to aspiring scholars, inadvertent human errors were also possible. However, by checking several versions (where several existed) of the same hadith for exactness in text, one can determine which one is the most reliable (*thiqa*) from the common source, although this does not guarantee the correctness of the hadith, that is, whether the Prophet had actually said it or not (see below, and also chapter 13).

Some prominent hadith transmitters were minor at the time of the Prophet's death, but many of their hadiths were accepted as genuine. There are also many contradictory hadiths, which Imam Shafi'i tried to solve by using a rule of abrogation, the later hadiths replacing the earlier ones on the same topic. But the validity of this abrogation policy is questionable, since there is no record that the Prophet had ever authorised such a policy for hadiths. In addition one must ask how Imam Shafi'i and his followers determined the time of the hadiths to ascertain its context, given the presence of so many other uncertainties, including missing transmitters in the *isnads*. Some hadiths contradicted each other, some contradicted reason and some contradicted the Quran. Some of these rejected hadiths were included in later collections. Furthermore, there appears to be some evidence that even Imam Shafi'i, let alone others, did not apply that abrogation policy consistently.

There is another aspect of the *isnads*, which relates to continuity. The relevant terms are:

musnad: no break in the chain of transmitters

muttasil: the connection among the transmitters are known to be feasible, but may have stopped before reaching the Prophet.

marfu: the connection goes right to the Prophet.

mursal: isnad with generation gap between two successive transmitters.

The best hadiths are *marfu, muttasil, musnad*. The *isnad* is broken if a connection between two successive transmitters in the *isnad* is not known, or if some transmitters are not known to be reliable. It would appear that some hadiths with broken *isnads* had been accepted, which later led to controversies. Examples of *isnads* with generation gaps (*mursal*) are hadiths from Said bin Musayyab (634-714, Medina), Ibrahim Nakhai (670-714, Kufa), Hasan al-Basri (641-728) and Mohammad Sirin (653-728). Even though the first two died some 80 years, and the second two nearly 100 years, after the death of the Prophet (d 632), their hadiths without any link to the Prophet in the *isnads* had been accepted as valid and used in the formulation of Sharia. There are many examples of this kind.

It would also seem that *isnad* did not exist in the first century of Islam. Muhammad bin Shihab al-Zuhri (d 741 CE, 124 AH), the father of hadith studies collected *isnads* at the start of the second century of Islam. According to his outstanding pupil Imam Malik Ibn Anas (715-795 CE), many of these *isnads* were found to be incomplete and imprecise, but were accepted as valid later in the second century of Islam. Malik, who was well-known for his thorough scrutiny, analysed 1,720 hadiths, among which only 600 were traced back to the Prophet, others had partial or no *isnads*, and furthermore some of them were not hadiths of the Prophet, but from his Companions or later scholars. Another study of 1,005 hadiths was conducted by Shaybani, the great scholar (see later), of which only 429 were found to be originating from the Prophet. Most of the hadiths for which Malik could not find any *isnads*, appeared in later compilations with *isnads*! So how reliable are they?

Finally observe that the scrutiny technique described above relates only to the *isnads*, but not to the *matn*, that is, we do not know if the words in that *matn* are really from the Prophet. However the revered hadith compilers cited below did not even use these detailed *isnad* tests. They developed their own less rigorous checks and defined their own version of *sahih* hadiths, resulting in many versions of *sahih*, some being more *sahih* than others!

Hadith Compilation

Following the claim of Imam Shafi'i that the entire corpus of the sunna of the Prophet was available among the hadith experts, a serious effort was made to compile hadiths. The most famous among these compilers were Mohammad bin Ismail al-Bukhari (810-870) and al-Muslim bin Hajjaj (817-874, 203-261 AH). Each compilation contains about 4,000 *sahih* hadiths, but the term *sahih* has been defined slightly differently in the two compilations (see below). In al-Bukhari [Buk84] the same hadith is sometimes repeated under different categories, which makes the total number with repeats around 7,275. Al-Bukhari included hadiths from 208 and al-Muslim from 213 Companions of the Prophet, with 149 Companions in common. Furthermore, al-Bukhari accepted 434 transmitters not in al-Muslim, while al-Muslim cited 625 transmitters not in al-Bukhari. None claimed to be exclusive. However, many hadiths in the two collections are common, but al-Bukhari is usually regarded as stricter (more authentic) by Muslims, for reasons explained below.

Al-Bukhari insisted for evidence that any two consecutive persons in the *isnad* had actually met, while al-Muslim required the consecutive two persons only to be contemporaries. Denouncing al-Bukhari's strictness as an innovation (*bida'*, and hence *haram*), al-Muslim divided hadiths into three grades: first were the *sahih* hadiths, the second and third grades were weaker versions of *sahih*. He defined a hadith to be *sahih* if the persons in the *isnads* are free of all faults (that is, they are pious) and all the different versions are precisely similar in text. If the persons in the *isnad* were of implacable piety and if their memory is slightly impaired compared to that in the *sahih* grade, then the hadith was classed as second grade. If a hadith qualifies for second grade, except that the memory impairment is more serious, then it placed into the third grade. Others hadiths were excluded. However, these criteria were very subjective: what is good memory, what is precision and what is piety? Furthermore, piety is not always necessarily a guarantor of truth. There were criticisms of both the compilations on these points when they were produced, but later on the criticisms gave way to reverence towards these two scholars. Subsequent scholars defined another type of *sahih* hadiths, in terms of al-Bukhari and al-Muslim as follows, with decreasing *sahih*-ness

1. found in both
2. found in al-Bukhari alone
3. found in al-Muslim alone
4. any hadith outside al-Bukhari and al-Muslim, but match the criteria of the both
5. any hadith outside al-Bukhari and al-Muslim, but match the criteria of one of them.

So there are many different definitions of *sahih*. Finally these two compilations, al-Bukhari and al-Muslim, are also called *sahih* works, to distinguish them from the other compilations, of which two well-known ones are by:

(1) Abu Daud (817-88) who had studied under Imam Ahmad Hanbal. His collection of about 4800 hadiths classed under his version of *sahih* (sound), *hasan* and even *da'if* (weak) hadiths. His

definition of *sahih* was not as rigorous as, that of al-Bukhari or even al-Muslim.

(2) Al-Tirmidhi (821-92), a student of both Abu Da'ud and Bukhari, produced another collection, based on his subjective classification, such as *sahih, hasan, hasan sahih*, etc.

There are two further collections, one by al-Nasai (d 916) and the other by Ibn Maja (d 886). Therefore, there are altogether six collections of hadiths.

However some Muslims were not convinced with the rigour of the authenticity from the very beginning, and therefore they came to reject all hadiths as being not genuine, the Mutazilites among them. Western scholars who studied hadiths tend to agree with these sceptics, even though the Muslim ulama never doubt the validity of the hadiths, even when some glaring contractions are present (see chapter 13).

There were two earlier special compilations, one by Abu Da'ud al-Tayalisi (d 819) and the other by Imam Ahmad Ibn Hanbal (d 855). These special compilations were called *musnads* as they arrange hadiths according to the transmitters (in the *isnads*) rather than according to topics. Al-Tayalisi collected 2767 hadiths from 281 Companions of the Prophet, while Ahmad Hanbal collected 30,000 hadiths from some 700 Companions. But Ahmad Hanbal accepted not only weak but also fabricated hadiths in his collection. As he became a hero of the orthodox, after 16 years in prison, his collection became very popular among them.

Need for Further Validations

The validation of hadiths was based on *isnads* (which is an external criteria), rather than *matn* or contents (internal criteria). In addition, one must also consider context. We shall comment on these points briefly below, but not necessarily in that order.

As discussed earlier, a set of criteria had been developed to verify *isnads*, such as the piety of the narrator, memory, and so on. Piety can only be directly tested (if at all) on the last person of the *isnad*, since the other persons on the *isnad* would have been dead. But is piety a guarantor of accuracy? On correctness of the collected hadiths, al-Bukhari applied *isnad* criteria most strictly, but beyond checking that the two successive persons in the *isnad* met, even he did not (and could not) check that the transmission was correct. Often (there are examples) a person heard a hadith from many people (perhaps slightly different versions), for which he mentioned only one *isnad*, with one common version of the hadith. Also a reliable person could have reported a hadith from an unreliable source (predecessor), but that unreliability was not recorded although the name of the source was.

If this *isnad* verification process of al-Bukhari was not strict enough, the other compilers have employed even less strict criteria, almost invariably with subjective judgements. Some hadiths were accepted (even by Shafi'i) because the content looked right, although the *isnad* was weak. Sometimes missing parts in an *isnad* were completed subsequently with plausible information from other sources, which meant that if the hadith was in error to start with, the erroneous hadith was fortified wrongly. It is surprising that instead of admitting with humility the weakness of their verification process, some revered hadith compilers like al-Muslim accused those, who applied stricter criteria, of committing the sin of *bida'*! It seems that the hadith compilers did not think it necessary that they should apply utmost rigour to discover the true hadiths.

An associated consideration is the context of a hadith. What the Prophet said in a particular context, at the instigation of a particular event, may not have been meant to hold in other contexts and time. Such a study, which has not been made, could not only provide an understanding of some contradictory hadiths, but could also make the interpretation of many other hadiths more accurate.

The final point is the content criteria, that is, whether the content is plausible. This also needs investigation. A statistical cryptographic word analysis technique (chapter 5) had been developed by Kindi (d 870 CE) principally for this purpose, but was never applied to determine the etymology of words, the structure of sentences and linguistic patterns in hadiths. Modern versions of similar techniques have been employed successfully to the Bible for text analysis. In fact some 20th century Muslim scholars had been declared heretic just for contemplating the idea of such an analysis (chapter 13). Thus, reasoning has been deliberately banished from the hadith criteria by the orthodoxy, but according to many Muslim scholars the time has come to re-examine these issues (chapter 13).

4.2 Sharia Law

It has been claimed [Sch64] that the sunna of the Prophet did not exist as a separate body of knowledge until about the middle of the 2nd century AH (8th century CE). There was instead the sunna of the Community, different communities, such as the Medinan or the Kufan school, having different sunnas, which Fazlur Rahman [Rah79] called the "living tradition". It is assumed that these sunnas were based on the Quran and the traditions of the Prophet as understood, interpreted and applied in different new situations, within the context of the local practices. What Imam Muhammad Abu Idris Shafi'i (767-820) attempted to do was to produce a set of universal laws, purely based on the Quran and hadiths (for sunna), independent of the local differences (for example, the Kufan school was more liberal in its outlook than the Medinan school).

Earlier Opposition to Hadiths as the Basis for Law

It was not only the Mutazilites, but also the lawyers and jurists who objected to hadiths, although the reasons were different. The lawyers and jurists accepted hadiths in principle, but found them unreliable and unrigorous for application as a basis of law, while the Mutazilites took a stronger position rejecting hadiths totally because of their unreliability. The Mutazilites were not only theologians, but they were also lawyers and jurists. Their basic argument for rejecting hadiths "was essentially the same as that of other legal schools", which was "that since its essence is transmission by individuals, [it] cannot be a sure avenue of our knowledge about the Prophetic teaching unlike the Quran about whose transmission there is a universal unanimity among the Muslims" [Rah79/p62]. They cited many contradictory hadiths and over-eager attitude of the hadith collectors to accept even the dubious ones because they sounded right. Is it proper, the Mutazilites asked, to dissect the Quranic injunctions on the basis of such unreliable hadiths? While rejecting hadiths as unreliable, they nevertheless accepted sunna and *ijma'* (see below) for the interpretation of the Quran.

In addition, Malik Ibn Anas and his contemporaries did not believe the hadiths to have any over-riding authority, unless they were in accord with the practice of a community [Gui87/p99,p95]. If one accepts the thesis of Imam Abu Hanifa (d 767) and other jurists that only a very few hadiths are genuine and that there are not enough genuine hadiths to build a law, then the whole foundation of Sharia becomes vulnerable. While during their lifetimes, both Abu Hanifa and Malik were able to place reason and local *ijma'* as alternative to sunna, it was no longer easy to do so after Imam Shafi'i. Obviously Islam became more conservative with time.

Earlier Practices

The Prophet made legal decisions based on the Quran and the current practices, which were largely tribal, treating the umma as a kind of super-tribe. The first four Khalifas (Rashidun) continued to follow in general the Prophet's precepts, but sometimes deviating from them. Khalifa Umar in particular took a liberal interpretation of some Quranic injunctions, despite objections from the orthodox; for instance by not applying the Quranic punishment for theft when someone had to steal because of extreme hunger in a famine [Zak89/p52]. Another clear and very significant case was his decision not to distribute conquered lands as war booty to the soldiers, ignoring the Prophetic tradition. The following two Quranic statements are relevant here:

".. All spoils (of war) belong to God and the Apostle ... ".
Quran [08:01].

"And know that whatever booty that you may acquire (in a war), one-fifth thereof belongs to God and the Apostle, and the near relatives, the orphans, the needy, and the wayfarers; ... ".
Quran [08:41].

These two verses imply that the spoils of war should be distributed as the Prophet wishes, confirming at the same time that the Prophet's own share should be one-fifth as he had to provide for other deserving groups. According to some scholars (see Muhammad Assad's translation of the Quran), one-fifth was the minimal amount; the state (i.e the Prophet) could have taken the whole amount by virtue of the first verse above, but the Prophetic tradition nevertheless was to distribute the four-fifths to the forces.

However, as stated above, in spite of the Prophetic tradition, Khalifa Umar declared all conquered lands (war booty) to be the property of the state (without any share of the soldiers in it), against much protestation at the time, although eventually everyone recognised the wisdom of Umar. The argument Umar used was the wider context of the new situation as against the limited Arab context of the Prophetic example. This principle of state ownership later became the Muslim tradition, even when it harmed the inhabitants. The later Muslim rulers applied this principle to grab all lands for themselves (the state), which prevented the formation of landed gentry (see also chapter 9 for its adverse impact). No other Khalifa, after Sharia law had been fixed in stone, was allowed to violate the Prophetic traditions captured in Sharia, say, by using Umar's argument of context[4], to derive new solutions for new situations.

It should also be mentioned that sometimes Umar did not follow the Quranic provisions and gave harsher punishment as in the case of adultery. This did not cause too much problem, as the idea of harsher punishment for sin was popular. However, it should be admitted that Khalifa Umar, the architect of the Islamic state, not only went against some Prophetic traditions, but also did not always adhere literally to some Quranic provisions as mentioned above. Therefore one might legitimately ask: if such changes were required within a few years after the death of the Prophet, how many more changes could be needed after 1400 years? We shall return to this question and revisit the points raised in chapter 13.

Not only Umar, but also the other members of the Rashidun, in some cases, had to add new features based on needs and pragmatism – sometimes they even reversed their own judgements on appeal (as Umar did), without giving any higher principle on which the reversal was based. Khalifa Ali refused to be bound by the tradition of Abu Bakr and Umar, presumably because he thought they did not always follow the Prophet sufficiently faithfully. However, as the Rashidun, the first four Khalifas were considered the legitimate highest authority both in religion and administration, and hence their decisions were accepted as models. But the fact remains, that there had never been any higher code in Islamic law, such as the code of Hammurabi in ancient Babylon, to serve as guiding principles. The Quran has only about 600 ayahs (out of a total of some 6200 ayahs) which deal

---

[4] Some further examples of Umar's deviation from the Quran and hadiths can be seen in [Kam99/p139].

with rules and regulations, most are on religious rites and duties, and only around 80 ayahs concern what we might call, social law, as discussed here.

Under the Umayyads, as the empire expanded, new problems and conflicts arose which required new laws, as stated earlier. They adapted the local Arab tribal laws, made non-Arab Muslims second class citizens[5], but otherwise followed pragmatically the local customs, thus creating different local sunnas. They created the office of qadi as a judge. Apart from being subject to a central administrative control, the qadis remained autonomous in their judgement on actual cases, sometimes refusing to accept any intervention from the Khalifa himself. It was possible to have entirely opposite judgements on two similar cases depending on the judge, as pointed out earlier.

The Abbasid Khalifa Mansur was advised to use his authority to enact unified laws, independent of the religious orthodoxy, for the empire in 757 CE, but he instead deferred his right to the ulama, who were then largely concentrated in two schools of law, one at Kufa dominated by Imam Muhammad Abu Hanifa (d 767) and the other at Medina dominated by Imam Malik Ibn Anas (d 795), both drew heavily from local customs. Some 60 years later when Khalifa Mamun wanted to moderate the trend in law making with the Mutazilite doctrine of *Khalq-i-Quran*, it was too late. While the Mutazilites championed human reasoning as the basis of law, the orthodox did not believe in reason at all, culminating in a power struggle in which the orthodox eventually won.

Legal Theory of Imam Shafi'i

The pro-hadith scholars, who did not believe in human reason as a basis of law, nevertheless employed the abstract reasoning technique of the Mutazilites to analyse and synthesize the Quranic text and sunna to create law, carefully shunning the Mutazilite interpretation. The master architect of Islamic law was Muhammad Abu Idris Shafi'i (767-820), an intellectual giant originally a member of the Maliki School in Medina. He defined clearly and coherently the four sources of Islamic law as:

---

[5] The ulama and the non-Arab Muslims (particularly the Persians) accused the Umayyads of discrimination against them.

Quran
Sunna
Ijma' (consensus of the umma)
Ijtihad (Qiyas)

A sunna, as stated earlier is the tradition of the Prophet. Therefore what he has asked his followers to do in his hadiths are sunnas, and so are his actions and recommendations reported by one or more of his Companions in their hadiths. Even though many Imams accepted sunna as providing a partial interpretation of the Quranic law, Shafi'i elevated it to the level of divine interpretation, and henceforth the only interpretation. Shafi'i did not believe that there could be any conflict between the Quran and sunnas (or hadiths). But if there were any apparent conflicts with a hadith, he put the hadith higher as the correct interpretation of the divine will, which might appear contradictory to the minds of the ordinary humans. Shafi'i interpreted the numerous Quranic injunctions, such as:

"... He sent down to you the Book and Wisdom for your Instruction. ..." Quran [02:231].

to follow wisdom, as an injunction to follow the actions of the Prophet alone, but not to take any actions not taken by the Prophet. This perhaps explained why Imam Hanbal refused to eat watermelon, in the absence of any explicit evidence that the Prophet ever ate it. Shafi'i held the position that a sunna can be abrogated by only a stronger sunna, but *not* by the Quran, even though there is no hadith to support this assertion. Since hadith became the central source, its correctness became important.

The concept of *ijma'* is based on the following hadith:

"My community (umma) will never agree on an error"

Imam Shafi'i defined *ijma'* as the consensus of the whole Muslim umma, but not of a locality, as hitherto practiced by the Hanafi and the Maliki schools. *Ijma'* cannot contradict the Quran or hadiths.

The fourth source of law was defined as *ijtihad* (personal intellectual struggle), a particular form of *ijtihad* being the reasoning by analogy or *qiyas*. *Ijtihad* can be applied only when it does not violate the other three sources, but subsequently the scope of *ijtihad* had been narrowed down first only to *qiyas* and then to a special form of *qiyas*, devoid of any *ra'y* (considered personal opinion) and especially denuded of any personal discretion (*istihsan*). By the year 900 CE, Imam Shafi'i's ideas had been largely accepted as the basis of Islamic law.

Four Schools of Law

Eventually the ulama settled on four main Sunni schools of law (*madhabs*, to be pronounced as mādh-hābs): Hanafi, Maliki, Shafi'i and Hanbali with equal validity, all heavily influenced by the work of Imam Shafi'i.

Khalifa Umar placed a garrison in Kufa, which later produced the Kufan school of law centred around the Companion Abd Allah Ibn Masud (d 32 AH/653 CE). His ideas were carried forward through the works of Hammad (d 738), Ibn Abu Layla (d 765) who was a qadi, and Abu Hanifa al-Numan Ibn Thabit (699-767). While Ibn Abu Layla developed a practical common sense approach, it was Abu Hanifa who produced a systematic theoretical approach to technical legal thoughts, and hence it was later named by his followers as the Hanafi school of law (madhab). Abu Hanifa, a former merchant from Kufa and a convert, pioneered the new discipline of jurisprudence (fiqh). A jurist is a faqih. Abu Hanifa did not subscribe to the *as'hab al-hadith* (the partisans of the tradition) supporters, as he believed in *ijtihad* (independent reasoning) with the ability to make new laws.

The Hanafi school was more liberal and broad-minded, perhaps because of the Persian influence, and was particularly favoured by the Abbasids. Its doctrine was developed fully after the death of Imam Abu Hanifa, partly by Abu Yusuf (d 798) the chief qadi of Khalifa Harun al-Rashid, but mainly by their common and brilliant student Shaybani (d 805), originally from the Maliki school in Medina. It was apparently Shaybani who was the real creator of the Kufa school, under the name of his revered teacher Abu Hanifa.

Malik Ibn Anas (715-95) was a pupil of Muhammad Zuhri, who was a pupil of Urwa, a nephew of Ayesha – all of them were great hadith experts in Medina. Malik's two distinguished pupils were Shafi'i who formed his own school and Shabyani who joined the Hanafi school. Compared to the Hanafi's, the Maliki school was narrower in its outlook reflecting the relative conservatism of Medina.

The followers of the legal theory of Shafi'i established the Shafi'i school of law, which accepted any hadith that sounded right as valid, even if the *isnad* was faulty. The Hanafis were stricter in *isnad* verification, while both the Hanafis and the Malikis retained some of their local consensus of Kufa and Medina respectively. Imam Hanbal, collected 80,000 hadiths and established the Hanbali school of law based strictly on hadiths, rejecting both *ijma'* and *ijtihad*. A fifth school called Zahiri, based on the literal interpretation of the Quran and hadiths, was established by Dawud Ibn Khalaf (d 883). One of its later outstanding adherents was Ibn Hazm (d 1064) who denounced *qiyas* as a perversion and heresy, but his books were later burnt in Seville. The school disappeared shortly afterwards.

The Hanafis were supported both by the Abbasids and also by the Ottoman Turks, and as a result they are widespread now in Iraq, Turkey, Jordan, Syria, Palestine, Afghanistan and in the Indian Subcontinent. The Malikis are predominant in Africa (including Egypt) and in the past in Spain, the Hanbalis (Wahhabis) in Saudi Arabia and the Shafi'i's in Indonesia and Malaysia (exported by Arab merchants). From the 14th century onward the Hanbali school declined until revived by the puritanical Wahhabi movement, founded by Muhammad Ibn Abd al-Wahhab (d 1787), a follower of Ahmad Taqi al-Din Ibn Taymiyya of Damascus (1263-1328). Because of their intolerant attitude to fellow Muslims, the earlier Wahhabis were distrusted as good Muslims, until the recent times when the oil-rich Saudi Arabia (Guardian of the two Muslim holy places) adopted Hanbalism-Wahhabism as their madhab. But Wahhabism is most anti-rational and is regarded by many Muslims as extreme. Talibanism in Afghanistan is a manifestation of Wahhabism. They even reject the Abbasid Golden Age of Islam as being un-Islamic!

The Shias accepted hadiths and sunna, but only those hadiths that were transmitted by their Imams. *Ijma'* and *ijtihad* were restricted to their Imams and ayatollahs. It may be stated that the Abbasid and later Khalifas kept some law-making power outside Sharia under what is called administrative law (*qanun*). That included police, criminal justice, and laws on taxations and properties (not inheritance). In this sense two systems of law co-existed: Sharia (divine) and *qanun* (human). Some later Ottoman Khalifas tried to extend the jurisdiction of the qanun at the cost of Sharia, with inevitable clashes with the ulama.

4.3 Limitations and Consequences

As time passed, *ijma'* soon became defined in a restrictive manner as the collective agreement of the *qualified legal scholars* in a given generation. However, different *ijma*'s produced different agreements, thus generating conflicts. To avoid future conflicts, every previous decision of *ijma'* became infallible, except for a repeal by another *ijma'* in a later generation, which was highly unlikely in practice. Consequently, the laws derived through *ijma'* became increasingly rigid. In the early 10th century, *ijtihad* became a spent creative force within the restrictions of *ijma'* and within other restrictions that were imposed on the nature of reasoning. Although no one formally declared it, the gate of *ijtihad* was deemed closed in the 10th century, and the right of *ijtihad* was replaced by a new doctrine of the duty of *taqlid* (imitation). Henceforth every jurist was an imitator, bound to accept and follow the doctrine of the predecessors, rather than taking recourse to *ijtihad* to produce a new solution. This restriction, although advocated in the 10th century, was really adopted after the Mongol invasion in the 13th century, returning in a sense more to the past. This has institutionalised an exaggerated respect for the personalities of the past, placing their deeds and opinions beyond any scrutiny, and thus fossilising Islamic law.

The closure of *ijtihad* did not prevent new legal judgements, what it stopped was the addition of new legal principles. As a result, concepts, such as an institution as a legal entity, personal liability, negligence, and the equality before law are unknown to Sharia law. Equally unknown are any rules for evidence (just see how evidence is collected and presented in some recent well-publicised cases involving foreigners in Saudi law courts), the concept of innocence unless proven guilty or anything else that today we would place under human rights. Bribery or financial corruptions are not handled satisfactorily within Sharia law. For example, in stealing from the state (Baitul Mal), one is taking from a fund which is partly one's own, and hence it is not as serious a crime as stealing from a person [Gru77/p491]. Under Sharia, there is no central court, no autonomy of judges, nor any institutional mechanism to take previous judicial rulings into account.

The idea of synthesis was unknown in Islamic law. After Shafi'i it was not possible for anyone to come out with a new, say higher, principle of law, which subsumes other laws and which perhaps points out loopholes in the existing laws. All doors were shut on any developments based on human rational reason. This contrasts with the Mutazilites who wanted reason to be given back to the human beings, who could decide through freewill what was good and what was evil, who could act (create their own acts) and who could be administered the necessary punishments as required for justice. Shafi'i's doctrine has led to the complete subordination of human reason to the understanding of the divine law, in fact it outlawed human reason as a source of law. Most importantly, it discouraged rational reasoning in the society, which is the basis of all developments in science.

The Muslim scholars of the recent centuries are all dubious of the reliability of hadiths in those six collections and of the validity of Sharia law based on them. The inability of Muslims to meet the challenges of the modern world has produced two opposing trends among the scholars: *ahl al-hadith* scholars who support hadiths in principle, but are interested to re-study hadiths (and also re-examine their authenticity), and *ahl al-Quran* scholars who believe the Quran alone to be the complete source and hence the hadiths to be dispensable. Some of their reasons and ideas will be aired in chapter 13 of this book. None of these groups believes in the current Sharia law to be right, some even subscribe to the separation of state and religion.

References and Sources

[Abb85] Nabia Abbott: Two Queens of Baghdad, Saqi Books, London 1985.

[Ada76] C. C Adams: The Authority of Prophetic Hadith in the eyes of some Modern Muslims, edited by D. P. Little, Essays on Islamic Civilisation, E.J. Brill, Leiden, 1976.

[Ahm97] Kassem Ahmad: Hadith – A Revaluation, published by Universal Unity, ISBN 188 189 3022, Paperback, 1997. Also available in the Internet: www. free-minds.org. His five lectures on this topic in a Malaysian University was published as a book in 1987, later banned, and then translated into English with some improvements and published in 1997. This is worth reading.

[Arm01] K. Armstrong: Islam – A Short History, Phoenix, 2001.

[Buk84] Al-Bukhari: Hadiths, trans by M. M. Khan in 9 vols, Kitab Bhavan (India), 1984. [Also available in the Internet]

[Bur94] John Burton: An Introduction to the Hadith, Cambridge University Press, 1994.

[Cou64] N. J. Coulson: A History of Islamic Law, Edinburgh University Press, 1964.

[Gui87] Alfred Guillaume, Islam, Penguin, reprinted 1987.

[Gru77] G. Von Grunebaum: Islamic Society and Civilisation, in Cambridge History of Islam, ed: P. M. Holt et al, Vol 2B, 1977.

[Juy83] G. H. A. Juynboll: Muslim Tradition, Cambridge University Press 1983.

[Kam99] Mohammad H. Kamali: "Law and Society", Oxford History of Islam, ed by J. Esposito, 1999. Chapter 3, pp107-54.

[Kam04] Mohammad H. Kamali: Hadith Studies, Islamic Foundation, England, 2004.

[Rah79] Fazlur Rahman: Islam, University of Chicago Press, 2nd ed. 1979. See footnote 8 in chapter 13 on Fazlur Rahman.

[Rut91] Malise Ruthven: Islam in the World, Penguin 1991.

[Sch64] Joseph Schacht: The origin of Muhammadan Jurisprudence, Oxford 1950, reprinted 1964.

[Sid61] M. Z. Siddqi: Hadith Literature – its Origin, Development and Special Features, Calcutta University, 1961.

[Zak89] R. Zakaria, The Struggle Within Islam, Penguin, 1989.

PART II

RISE OF SCIENCE UNDER ISLAM

CHAPTER 5
MATHEMATICS

The developments in mathematics under Islam drew from diverse earlier sources: Babylonian, Egyptian, Persian, Greek and Indian – particularly the last two. In the eighth and ninth century CE, many Greek works in mathematics were translated into Arabic either directly from Greek or via Syriac. The translated Greek sources included the Elements of Euclid, Conics of Apollonius of Perga, Spherics of Theodosius of Tripoli, Almagest of Ptolemy, works of Archimedes and mathematics from the Greek scientific centre at Alexandria.

A link with India was formed when Sind (now in Pakistan) was conquered by the Umayyads in 711 CE. It would seem that an Indian scholar from Sind arrived at the court of Khalifa Mansur in 773 CE, with a copy of what was called *Shiddhanta* (meaning decision) probably written by Brahmagupta (b 598? CE), on the movements of the stars. It was translated into Arabic at the command of Mansur by Ibrahim al-Fazar (d 777?), his son Mohammad and also Yaqub bin Tariq (d 796?), under the title *Sindhind*. It is believed that the direct contact with Indian astronomy and mathematics (including the Indian decimal numerals) started from that period. This translation was followed by many others. Translations from the ancient Persian sources included astronomical tables, such as *Zij al-Shah*. (*zij* means astronomical tables). However the ancient Persian contributions to science cannot be easily differentiated from the Greek contributions, since we know of them mostly through the Greek writings, the original Pahlavi (the ancient Iranian language) sources being lost a long time ago.

In reviewing Arabic mathematics below, we shall present some of the highlights covering the development of the basic numbering systems, arithmetical operations, algebra for problem solving, geometry, trigonometry and cryptography.

5.1 Numbering Systems

There were three separate numbering systems developed under Islam, of which the most widely used one was based on the Babylonian sexadecimal scale, but expressed – paralleling the Greek astronomical tradition – in Arabic letters.

This system of using letters for numbers is known as *jummal* or *abjad*, and it has no separate symbol for the digit zero. The *abjad* scheme was used for integers, fractions being calculated on the base of 60. For example, the fraction 15/20 is represented as 45 parts in 60, breaking it down into proper fractions of 60 as [30/60 + 15/60], implying [½ + ¼].

The first reference to Indian arithmetic in the Islamic world was made in 662 CE by Bishop Severus Sebokht of Syria, but the earliest known Arabic book on this was written by Musa al-Khwarizmi (section 5.3). The recorded use of a symbol for the digit zero in India (as distinct from a symbol for an empty position) was found in an inscription in 876 CE[1] [Boy91/ p213]. The numeral system went through an evolutionary process and eventually settled down to two versions:

(i) Eastern Arabic 0 ١ ٢ ٣ ٤ ٥ ٦ ٧ ٨ ٩, where the symbol 0 for zero was later replaced by ・ (dot), and
(ii) Western Arabic (Andalusia): 0 1 2 3 4 5 6 7 8 9, from angled symbols as shown below:

These two versions evolved from the original Indian version over many years through many changes. The western version is said to be based on the idea of angles to represent digits, as shown in the diagram above: zero angle for the digit zero, one angle for the digit 1,

---

[1] A symbol for zero has been found in the records of temple accounts of the ancient Sumerians (4000 BCE), long before its partial use by the Babylonians. In India it was probably widely used long before 876 CE.

two angles for the digit 2, and so on. Observe how seven angles are used in digit 7 and nine angles in digit 9. It may also be observed that in Continental Europe (that is, excluding the UK and Eire) the digit 1 is usually written (in handwriting) even today with a sharp angle and digit 7 as given above, with an upward line ending in a cut in the middle. The current western version (i.e. the European form) seemed to have been fixed in the sixteenth century, probably in Germany [Boy91/p237]. However, in the Islamic world the decimal numbering system was used only haphazardly, the astronomers preferring the letter-number (*abjad*) scheme following the Greek astronomical tradition. There was a third scheme (called secretarial) in which the numbers were written in words, permitting some simple arithmetical operations on them.

[*Readers not interested in arithmetical calculation may prefer to skip sections 5.2 and 5.3*]

5.2 Arithmetical Operations

We sketch below the major developments on arithmetical operations, mainly in terms of the work carried out by Musa al-Khwarizmi, al-Uqlidisi, Jamshid Kashi, Umar Khayyam, Nasr al-Din al-Tusi, al-Kindi and some others.

The earliest work on Indian arithmetic available in Arabic is that of Ahmad bin Ibrahim al-Uqlidisi produced in 952-53 CE in Damascus. It was called *al-hisab al-hindi* and included new techniques for multiplication and division of decimal numerals. These techniques are fairly close to the modern techniques, though more laborious. For instance, in multiplication, they multiplied digit by digit separately, shifting each result to the left for the power of 10 as needed, and then adding these individual results – a laborious process. Uqlidisi also innovated a basis for handling the decimal fraction, separating it from the rest of the number by a ' / ' sign. Most importantly he provided a technique for finding square roots, as illustrated below.

The principle he employed for the evaluation of square root can be demonstrated in terms of three numbers N, p and q, where:

$N = (p + q)^2 = p^2 + 2pq + q^2$, and
$p > q$.

We wish to find the square root of N which is (p +q). In the evaluation process the value of q will be reduced (making the value of p larger) at each step until the value of q is zero. The process begins in step 1, with the selection of a value of p, say p', such that p'^2 is the largest number less than N. Then a value of q, say q', is selected as the largest number such that:

$$p'^2 + 2p'q' + q'^2 = N_1 \leq N.$$

In the second step, the values of p and q are revised as

p' = p' + q', where p' is now the new value of p.
q = q - q'.

and a value of this revised q, shown below as revised q' (different from the old q') is determined as the largest number such that:

$$p'^2 + 2p'q' + q'^2 = N_2 \leq N$$
where $N_2 > N_1$.

N_2 is the result of the second step. The process is repeated successively until the *i*th step where $N_i = N$. The latest (p' + q') yields the square root.

The process of evaluation with actual decimal numbers is made simpler by taking from the left each pair of digits in the number N at a time. If the number N does not have an even number of digits in it, then it is made so by prefixing it with a zero on the leftmost side. For example, the number 144 is made into 0144.

Each pair of digits contributes one digit to the root, from the left. We shall show below the principle of this process with a small number. We can choose the number 144, but to make it a little complex let us use a slightly larger number, say N = 6889, for which the leftmost pair is 68. First, we describe the process, which requires the successive determination of one number from each pair in N.

Determine the largest value m, such that $m^2 \leq 68$. (Note 10*m is equal to p'). Now find the largest value n (the same as q') such $(10m)^2 + 2(10m)n + n^2 = 100m^2 + 20mn + n^2 = N_1$ (where $N_1 \leq N$). The process continues successively for each successive pair in N from the left. More explicitly, for the above example m = 8, since its square 64 is just smaller than 68, the first pair of our number.

```
      6889
        64      [we are actually subtracting 6400, i.e. (10*8)²]
      ------
       489      [from which we subtract 2(10*8)n + n² for a suitable n]
       489      [2(10*8)n + n² for n = 3]
      ------
         0
```

So the answer is <m> <n> = <8> <3> = 83. That is, $10m + n = 83$.

If N has j even number of digits, then there will be $j/2 = k$ pairs. For each successive pair from the left, the revised p' has to be multiplied by 10 (as shown above for m) to take account of the decimal shift of one position in the resultant square root. The principle should be clear from the above example. This technique is in fact still the basic procedure taught in schools today, and is sometimes used in computer algorithms.

5.3 Algebra

The great Arabic mathematician Muhammad Ibn Musa al-Khwarizmi (d 847), who is widely accepted as the father of algebra, worked in Baghdad during the reign of Khalifa Mamun. Besides works on astronomical tables and astrolabes (see section 6.3), Khwarizmi wrote two books on arithmetic and algebra, one of which entitled: *Kitab al-mukhtasar fi hisab al-jabr wa'l muqabala* (roughly translated as *the book of summary concerning calculations by transposition and reduction*). Part of this book was translated into Latin in 1145 by Robert of Chester under the title: *Liber algebrae et almucabala*, thus coining the term algebra. Much of the Arabic draft, which included a section on Indian Reckoning, is missing. Khwarizmi's work on Indian Reckoning describes the Indian numerals (as the Hindu numerals) extensively with great elaboration, so much so that the Western scholars started calling them Arabic numerals, which he himself did not.

The Latin translation of his work discusses algebraic techniques which requires a problem to be expressed in one of these six forms:

(1) $ax^2 = bx$,   (2) $ax^2 = c$,   (3) $bx = c$,
(4) $ax^2 + bx = c$, (5) $ax^2 + c = bx$, (6) $bx + c = ax^2$.

Note that the forms (4), (5) and (6) can be expressed as a single equation if we allow the constants a, b and c to take both positive or negative values. In fact we can replace all six equations by a modern quadratic equation of the form $ax^2 + bx + c = 0$, where each of the constants a, b, c and also the variable x can take either a positive or negative value.

However, in the above six equations, each of a, b, c and x is always a positive integer. After transforming an equation into one of these six standard forms, it is solved geometrically with diagrams of squares and rectangles, the sides representing numerical values. Take for instance the quadratic equation: $x^2 + bx = c$, which we shall use below to explain that technique in terms of modern algebra:

$$x^2 + bx = c$$

or $x^2 + 4(b/4)x + 4(b/4)^2 = c + 4(b/4)^2$
[i.e. $4(b/4)^2$ is added on both sides]

or $[(x + 2(b/4)]^2 = c + 4(b/4)^2 = c + (b/2)^2$

or $[(x + 2(b/4)] = \sqrt{[c + (b/2)^2]}$

or $x = \sqrt{[c + (b/2)^2]} - (b/2)$

Observe that only the positive roots are taken. Let us take a concrete example, such as $x^2 + 5x = 66$. The value of x from the algebraic solution above, with b = 5 and c = 66:

$$x = \sqrt{(66 + 2.5^2)} - 5/2 = \sqrt{72.25} - 2.5 = 6.$$

This solution does not give the negative root, which however can be obtained from the modern general solution of the equation:

$ax^2 + bx + c = 0$,

where $x = b \pm \sqrt{(b^2 - 4ac)}]/(2a)$.

With a = 1, b = 5, c = −66, the value of $\sqrt{(b^2 - 4ac)}$ is 17,

and hence x = 6 or −11.

5: MATHEMATICS

So, what we have given above is the equivalent solution in a modern algebraic form. We shall now demonstrate the solution of Khwarizmi (figure 1), for which:

(i) We draw one large square, say square P, with sides of (x + b/2). At this stage, x is an arbitrary value (variable).
(ii) Then inside this square P, we draw concentrically a smaller square of x and call it square Q as shown in figure 1.

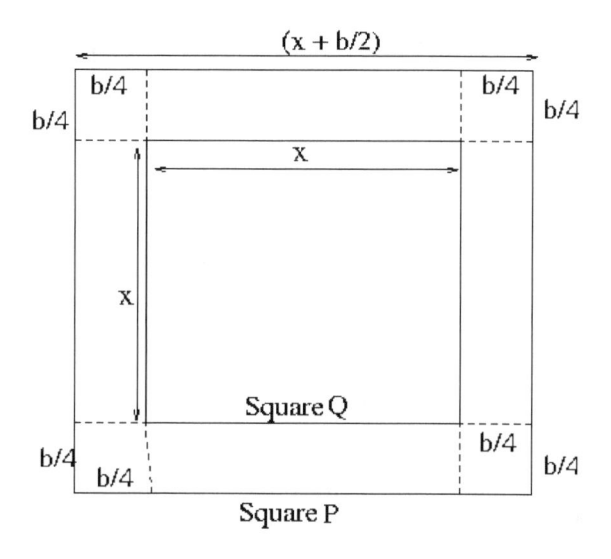

Figure 1: Squares for Algebraic Solution

Let us now consider the equation $x^2 + 5x = 66$ given earlier. If we use the value 5 for b, then the total area of the square P is

$$[(x + (b/2)]^2$$

$$= [x + 2.5]^2$$

$$= x^2 + 5x + 6.25$$

$$= 66 + 6.25, \quad \text{since } x^2 + 5x = 66$$

$$= 72.25$$

Each side of P is therefore

$(x + b/2) = \sqrt{72.25}$

or $(x + 2.5) = \sqrt{72.25} = 8.5$

Hence $x = 6$.

The book of Khwarizmi also contains rules on binomial expressions and a number of other ideas, which today would be placed under the *number theory* of mathematics.

Another great contributor to mathematics was Thabit Ibn Qurra [836-901 CE], who was responsible for translations of some major Greek works, including those of Euclid, Archimedes, Apollonius and Ptolemy – without such translations some of these Greek works would have been lost to us. In the process of translation however, Thabit became an expert, contributing to the *number theory* and also providing generalised proof for the Pythagorean theorem. He introduced the first known mathematical analysis of motion, which he then applied to heavenly bodies and in particular to the visibility of the lunar crescent, thus furthering the cause of astronomy. He also developed the concept of infinity in the *number theory*, in which a part is equal to the whole, an idea that was taken up later bravely by Galileo in the 17th century when the prevailing Christian dogma barred belief in infinity.

5.4 Geometry and Trigonometry

Starting from the Greek and Indian sources, the Arabic scientists worked on the science of geometry (*Ilm al-Handasa*) from the tenth to the fourteenth century, developing new techniques (especially for spherical geometry) partly for use in optics and astronomy. Notable contributors were Jawhari (9th century) and Nayrizi (d 900), Umar Khayyam (1050-1123) and Nasr al-Din al-Tusi (1201-74). Geometry was also used for land surveys and measurements, hydraulic works and other engineering.

The concept of sine and cosine was developed in India, but the Arabic scholars produced new trigonometric functions and solutions for both plane and spherical surfaces, in addition to establishing a number of advanced trigonometric relationships (for example, expressions involving sine and cosine). Khwarizmi wrote a treatise which produced Arabic tables for tangents. These have been used in the study of astronomy (spherical astronomy). According to T. E. Huff [Huf95/p50]

> "Trigonometry – an essential part of mathematics for astronomy – was invented by the Arabs".

E. S. Kennedy [Ken70/p337], the historian of science of that period also concludes that trigonometry, the study of the plane and spherical triangles [i.e. triangles on a spherical surface], was "essentially a creation of Arabic-writing scientists". Much of these developments eventually culminated in the outstanding contributions of Nasr al-Din al-Tusi, who worked on non-Euclidian geometry, trigonometry and planetary motion (see chapter 6).

5.5 Cryptography

Cryptography was well-known from the ancient time for sending secret messages using encryption and decryption techniques. The message to be sent is called *plaintext* and its encrypted version *ciphertext*. Julius Caesar employed a simple mechanism in which each letter in the alphabet (in the plaintext) is replaced by a later letter (in alphabetic order) in the ciphertext, like the example below:

| Plaintext | A B C D ... Y Z |
| Ciphertext | C D E F ... A B |

Here A in the plaintext will be replaced by C, B by D and so on. So the word CONGRATULATIONS in plaintext will become EQPITCVWNCVKQPU in the ciphertext. In this cipher, the displacement = 2, that is, each letter is replaced by the second letter from the right (in alphabetic order). Instead, one can use displacement of n, where n is the relative position of the ciphertext letter in the alphabet. This type of encryption is known as Caesar's cipher, but is very easy to crack.

As business transactions started to grow in the Islamic empire under the benign Abbasid rule, there was a need to despatch confidential messages, with better encryption. The person who stepped in to provide that service was Yaqub Ibn Ishaq al-Kindi (d 870) – an aristocratic Arab Muslim, mathematician, astronomer and above all a philosopher (see chapter 10), whose primary motivation in this work was scriptural analysis to distinguish the genuine from the fake. He was the author of some 290 books on astronomy, mathematics, medicine, linguistics and music. His greatest treatise on cryptology entitled: *A Manuscript on Deciphering Cryptographic Messages* was rediscovered in an Istanbul museum in 1987. It is a remarkable work, which described statistics of Arabic word occurrences, phonetics, syntax and cryptanalysis. He was suggesting the use of statistical word frequencies for breaking codes. Here are some paragraphs from that book:

> One way to solve an encrypted message, if we know the language, is to find a different plaintext of the same language long enough to fill one sheet or so, and then we count the occurrence of each letter. We call the most frequently used letter as the 'first' letter, the next most frequently occurring letter as the 'second' letter, and the following most occurring letter the 'third', so on, until we account for all the different letters in the plaintext sample.
>
> Then we look at the ciphertext we want to solve and we also clarify its symbols. We find the most occurring symbol and change it to the form of the 'first' letter of the plaintext sample, and the next most commonly occurring symbol is changed to the form of the 'second' letter, and the following most common symbol is changed to the form of the 'third' letter, and so on, until we account for all the symbols of the cryptogram we want to solve.

Observe that his explanation written nearly 1200 years ago is very lucid. In addition to letters, he also allowed symbols to be used in the ciphertext, and gave examples of solutions. The most common letter in Arabic is ا (alef for A in English), and the second most common is ل (lam for L in English). They together (right to left in Arabic) ا ل (al) means "the" which is the most common word in Arabic, as in *al-Kindi*. Equivalently in English the most commonly used letter is *e* (12%). Instead of letter count, one can use word count, but in any case the ciphertext has to be long enough for the statistical pattern to make sense.

This is the kind of technique that was used to decipher, the letters between Mary Queen of Scots and her supporter Anthony Babington, by Sir Frances Walsingham with the help of the code breaker Thomas Phelippes in 1586 CE for Queen Elizabeth I of England. Babington's secret messages were carelessly long which provided a large enough sample for the statistical analysis. Although codes today are far more complex, we have to admire the Arabic techniques of the 9th century, which remained useful even in the 16th century.

The most widely used symmetric encryption technique today is the one known as DES (Data Encryption Standard) based on IBM's Lucifer technique. Here the key k is 64 bits (0-63), including 8 bits for error checks (e.g. checksum). The functions E and D are very complex, involving many layers of permutations and also encryptions on encryptions, for example:

$$P \to C_1 \to C_2 \to C_3 \to C_n,$$

where C_n is the final ciphertext. It can be broken with sufficient computing power, although very hard. It allows 2^{55} possible keys to choose from at each layer.

There are however better techniques for use in the Internet communications today, but in all such techniques for secret code, it is still important, as it was in Babington's time, to have:

(a) a large number of possible keys,
(b) frequent changes to the keys and
(c) short messages per key.

5.6 Other Mathematical Works

Further work on algebra using the geometry of conic sections was carried out by Abu Jafar al-Khazin (d 970?), Ibn Haitham (965-1039), and Umar Khayyam (1050-1123). Umar Khayyam was a great mathematician, but more well-known in the West today for his philosophical poems *Rubayyat al Umar Khayyam*. His contributions to mathematics included higher degree algebraic equations (cube, higher powers and roots) and binomial numbers, which he systematised as a branch of mathematics called *ilm al-adad* in Arabic, and which forms part of the modern *number theory*. In it he studied the behaviour of numbers and classified them into different types. When Umar Khayyam died in 1123, the Arabic science was in the state of decline, but some work was still being carried out, such as on logarithms, which became commonplace in the Islamic world by the 13th century. John Napier rediscovered them around 1614.

The finest treatise on Arabic arithmetic was written in 1427 by Jamshid Ibn Masud al-Kashi who was a Persian working at Samarqand at the court of Sultan Ulugh Beg, a grandson of Timur (Ulugh Beg is also mentioned in chapter 6). Kashi's book *The Key to Arithmetic* was a comprehensive compendium for applications in mechanics, land surveys and theoretical astronomy. He calculated the nth root of sexadecimal numbers and the value of 2π (giving the ratio of circumference of a circle to its radius) up to 16 decimal places:

$$2\pi = 6.2831\ 8530\ 7179\ 5865$$

No other mathematician achieved this accuracy until very late 16th century. Here is a post-Kashi mnemonic to remember the value of π: "How I want a drink, alcoholic of course, after the heavy lectures involving quantum mechanics". If you are averse to alcoholic drinks, you have the choice of pepsicola or even dietpepsi. The secret is that the number of letters in each word in the mnemonic yields the value of π as 3.1415 9265 3589 79, in full agreement with Kashi's value.

Another of his contributions was a system for the use of decimal fractions, which was 200 years ahead of its time. According to E. S. Kennedy [Ken86/p580]: "Jamshid (Kashi's) computational algorism exhibited a feel for elegance, precision, and control, which had never been seen before, and which was not to be surpassed for a long time to come. ... Iran's scientific output, though weakening, may have maintained her in a leading position throughout the 15th century. Thereafter the lead passed to the West".

References and Sources

[Boy91] Carl B. Boyer: A History of Mathematics, 2nd ed. John Wiley & Sons, 1991.

[Fak83] Fakhry: "Philosophy and History": The Genius of Arab Civilization, J. R. Hayes (Ed), Second edition MIT Press, 1983.

[Huf95] T. E. Huff: The Rise of Early Modern Science, Cambridge University Press, 1995.

[Ken70] E.S. Kennedy: "The Arabic Heritage in the Exact Sciences", Al-Abhath, Vol 23, 1970.

[Ken75] E.S. Kennedy: "The Exact Sciences", Cambridge History of Iran, Vol 4, ch 10, pp375-395, 1975.

[Ken86] E.S. Kennedy: "The Exact Sciences in Timurid Iran", Cambridge History of Iran (1986), Vol 6, 1986, pp 568-581.

[Hil93] D. R. Hill: Islamic Science and Engineering, Edinburgh University Press, 1993.

[Sab76] A. I. Sabra: Ch 7: "The Scientific Enterprise", Islam in the Arab World, ed. B. Lewis, published by Alfred A. Knopf, New York, 1976, pp181-200. Abdulhamid Sabra is a great Egyptian scholar and an authority on Arabic sciences.

[Sart27] George Sarton: Introduction to the History of Science, Carnegie Institute of Washington, 1927. It comprises three huge volumes (I, II, III), covering the period from Homer to the 14th Century – the size is equivalent to some 5500 A4 pages. Most scientists (including Arabic, Chinese, Indian and Japanese) are individually listed and contributions described. These are indeed the books that made Arabic contributions to science well-known.

[Sin00] Simon Singh: The Code Book, Fourth Estate, 1999.

[Sin96] Charles Singer: A History of Scientific Ideas, Barnes & Nobles, 1996, originally published in 1900 under a slightly different title.

[Tur97] Howard R. Turner: Science in Medieval Islam, University of Texas Press, 1997, ISBN 029278 1490.

CHAPTER 6

ASTRONOMY

Islamic astronomy was inspired by religious requirements to set the times of prayers, to find the direction of Mecca (*qibla*) and to determine the start and end dates of Ramadan, the start and end times of fasting (during Ramadan) and the dates of other religious events. Further motivation was provided by the Quranic encouragement to appreciate the glory of God by examining celestial objects. The earliest sources in astronomy to which the Islamic scholars had access were similar to those in mathematics, that is, texts from Babylon, Persia, Greece and India. Work started in the 8th century, when astronomical tables called *zijs* were translated into Arabic from Indian and Afghani texts. One of those tables called *Zij al-Sindhind* was compiled by Khwarizmi, but now lost except for some fragments that are available in the Latin translation of the revision carried out by Majriti (c 1000) in Cordoba. The most well-known Greek text translated into Arabic was the *Almagest* of Ptolemy, which included many important astronomical tables. The other significant Greek works translated into Arabic were: Ptolemy's *Planetary Hypotheses*, Theon's *Handy tables* and a number of treatises on *astrolabes*[1] (see also section 6.3). The most productive period of Islamic astronomy spanned from 825 CE to 1375 CE, after which began a fast downward slide.

Khalifa Mamun, who was a champion of astronomy, established an observatory in Baghdad. Astronomer Abu Abdullah al-Batani (858-929), son of a scientist in Harran and a convert to Islam, calculated the length of the solar year to be 365 days, 5 hours, 46 minutes, with a difference of three minutes from the current value of 365 days

---

[1] An astrolabe is an instrument (used before the invention of sextant), to observe the position of celestial bodies.

5 hours 49 minutes. He was one of the first persons to develop and to apply trigonometry on solid spheres for astronomy. The Islamic astronomers produced a rich corpus of literature, of which some 10,000 manuscripts have survived, but only a small fraction of them have so far been translated from the original Arabic. Therefore we do not as yet have the full picture of the extent of their achievements, but what we know so far unveils an era of flourishing research and developments in astronomy. In this chapter we intend to give only a flavour of the works they carried out, in order to demonstrate the intellectual and technical heights they reached. For this, we shall focus on two areas: (i) spherical astronomy and (ii) the planetary model, and conclude the chapter with a brief note on astronomical instruments.

6.1 Spherical Astronomy

Much of the work was based on the Ptolemy's model of a geocentric universe, which the Arabic scholars followed and improved, and in the process they produced the new science of trigonometry.

Note for the readers: *Readers, not interested in the mathematics of spherical astronomy, may prefer to skip the next subsection and the equations given in the later subsections.*

Spherical Coordinates

To discuss some of their works, it is important to be familiar with the spherical coordinate system. For this, imagine you are standing in position P on the surface of earth somewhere in the northern hemisphere, east of London (London is on the zero longitude). Your meridian is the half-circle passing through you (position P) and the poles. This is also your longitude (see figure 1), which is expressed as angular distance (angle λ here) from the zero longitude L of London in the diagram. Thus the angle BND (on the surface of the earth) at the north pole N, and equally the angle BCD at the centre of the earth C on the equatorial plane E (shaded) yields the longitude. The line CP gives your vertical direction from the earth's centre. The angle DCP (denoted as angle Φ) at centre of the earth, but vertical to the equatorial plane, is your latitude.

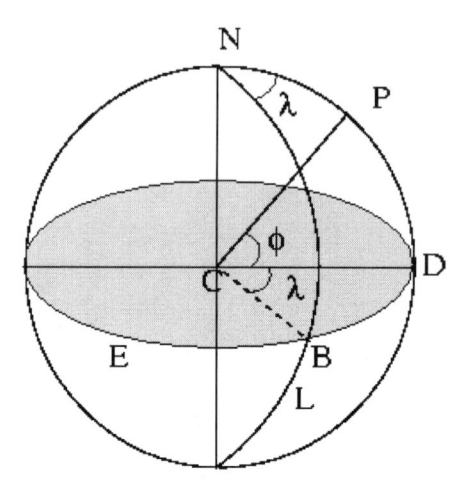

Figure 1: Longitude and Latitude

Now see the next diagram (figure 2) from which we have removed the longitudinal details for simplicity, but added the horizontal plane H (shaded), perpendicular to the vertical line CP. The angle DCB between the equatorial plane E and the horizontal plane H is the angle of inclination δ. *The altitude* of an object in the sky is the height from the horizon, expressed as an angle which can vary from 90 degrees (if above the horizon) to -90 degrees (if below the horizon), e.g. the altitude of an object O in the sky is shown as angle OCB given by α. Your latitude Φ (also in figure 1) is the same as the altitude of the pole star (in the direction from C to N) from your horizon. Thus the sum of angles δ and $\Phi = \pi/2$ radians (i.e. a right angle).

Associated with the *altitude* is the *azimuth* of a celestial object, which gives the angular distance of the object at the observer's (i.e. yours) horizon with respect to the north point of that horizon. To define it, imagine a vertical circle in the sky, going high over your head and passing through the celestial object. That circle will intersect your horizon at two points. The angular distance from the north point (of the horizon) to that intersection-point, measured in the eastward direction is the azimuth of that object. The azimuth can vary from 0 to 360 degrees (2π radians). The azimuth of the pole star is 0 degree. The parameters an observer at P can measure about a celestial object O are its altitude and azimuth, both with respect to the observer's horizon. These two parameters can be related by trigonometry and employed to derive other parameters of a celestial object.

6: ASTRONOMY

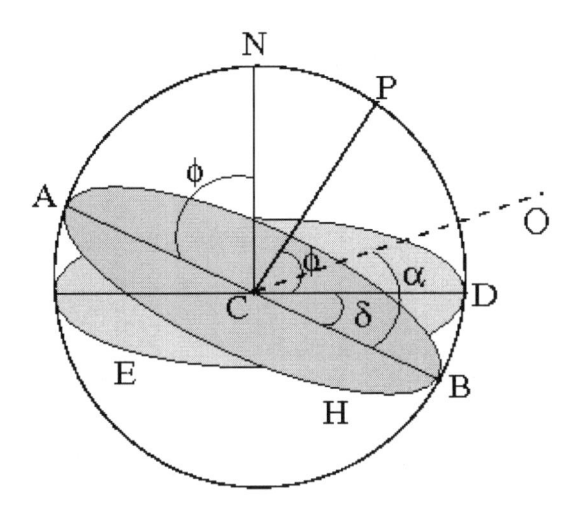

Figure 2: Altitude, Azimuth, Declination and Right Ascension

Similar to a position on the surface of the earth, we can define latitude and longitude of any object in the sky, if we imagine the sky to form a celestial sphere with all the celestial bodies (stars, the sun, planets and the moon) lying on the surface of that sphere, irrespective of their relative distances from us. In this celestial globe (sphere) the pole star will be the north pole, and the centre of the earth C the centre, with all celestial objects lying at equidistant from that centre C.

The equatorial plane of the earth will also be assumed to be the plane of the celestial equator. The latitude of a celestial object is called *declination*, and the longitude the *right ascension*. Therefore the latitude of the object O is the angle OCD which is denoted as angle (α - δ), and the latitude of the pole star is $\pi/2$ radians (i.e. 90 degrees).

The path the earth follows in its movement during a year is its elliptical orbit (these days we know it as the earth's orbit around the sun). The plane of this elliptical orbit makes an angle $23\frac{1}{2}^0$ to the plane of the celestial equator (which is the same as the earth's equatorial plane). Twice a year the two planes cross: spring and autumn equinoxes. The cross-point of the spring equinox, usually denoted as point γ (not shown) is taken as the zero celestial longitude (right ascension). The right ascension for a star is equivalent to the angle between its meridian and that of the spring equinox measured eastwards (as in the case of terrestrial longitudes).

Determination of the Position of the Sun for Prayer Times

These coordinates vary continuously for the sun (less speedily for the stars) throughout the year. At any given time t, the position of the sun is defined by its hour angle τ, which is given by the angle made by the plane of the object's meridian with that of the observer's meridian, which is given as:

$$\text{Angle } \tau = (\sin \alpha - \sin\rho \sin\Phi)/(\cos\rho \cos\theta)$$

where α is the altitude, ρ is the solar declination, Φ is the latitude of the observer, and θ is the solar right ascension. Both ρ and θ vary over time continually for the sun as stated above. This is a modern formula. In contrast, the Arabic scientists had a system of formulae that when evaluated produce the same result.

Determination of the Qibla

The determination of the qibla from any point on earth is a complicated calculation, where the essential challenge is the determination of a great circle passing through the coordinates of the observer and Mecca. The calculation involves many angles in trigonometry on a spherical surface. The Arabics had to develop spherical trigonometry for such calculations. A modern formula for it can be described as follows.

The direction q of the qibla from the point P (Φ, λ) of the observer to the point M (Φ_m, λ_m) of Mecca, where the pairs Φ and λ are their respective latitudes and longitudes, is given as:

$$\cot q = [(\sin\Phi \cos\beta - \cos\Phi \tan\Phi_m)/\sin\beta]$$
$$\text{or } q = \cot^{-1}[(\sin\Phi \cos\beta - \cos\Phi \tan\Phi_m)/\sin\beta]$$

where $\beta = |\lambda - \lambda_m|$ is the longitudinal difference between the two points.

In fact q is the angle made by the latitude Φ of P with the great circle through P and M. The Arabics did not produce this modern formulation, but an equivalent set, which consisted of a number of complex trigonometric formulae requiring a non-trivial calculation.

The qibla problem was solved mathematically by Nayrizi (Baghdad, c 900) and al-Biruni (973-1048) of Ghazna among others. There were also non-rigorous solutions based on astronomical tables for various parameters. Either way the correct direction of the qibla had to be determined each time a new mosque was built in the Islamic empire.

Abu Raihan al-Biruni of Ghazna studied mathematics (including trigonometry), optics, physics (specific gravity), astronomy (particularly astronomical geography) and pharmacology (covering over 700 drugs). He was recruited in 1018 CE, possibly under duress, by Sultan Mahmud of Ghazna (who ruled from 997 until his death in 1030) to accompany him on his annual raids to India, in order to write about India. Al-Biruni learnt Sanskrit, studied Indian arts and literature, folklore, religion, mythology and philosophy. According to Kennedy [Ken73] al-Biruni had written 148 treatises spreading over 13,000 folios (equivalent to 13,000 modern printed pages). No more than a fifth of his works have survived, but they all bear testimony to a great scientific mind, expressing his deep knowledge with precision and also great humour. He was indeed one of the greatest minds of all time, both in breadth of knowledge and in intellectual depth.

Al-Biruni was aware of the concept of a heliocentric planetary system (the ancient Greeks first thought about it – see chapter 1), but he never pursued it. Why? Was it because of the fear of the orthodoxy? We do not know except that al-Biruni always kept himself on the right side of the orthodox – he even despicably condemned Razi (see chapter 10) in favour of the orthodox.

Crescent Visibility

The third problem of spherical astronomy was the visibility of the lunar crescent. In theory a new moon is visible in the western sky by naked eyes if there is a difference of at least 12 equatorial degrees (48 minutes of time) between the setting of the sun and that of the moon. If the interval is less, the sky will not be dark enough for the moon to be visible to the naked eyes. The interval between these two setting times depends on three factors:

(i) the longitudes of the sun and moon, and their differences,
(ii) the latitude of the moon, and
(iii) the local terrestrial latitude.

The Arabic astronomers prepared tables, mainly for the setting intervals of 12 degrees for a range of values of the parameters needed for the calculation. Habash al-Marwazi spent a lifetime in astronomy observing lunar and solar eclipses. His calculations were verified by the correct prediction of the new moon of Ramadan in 860 CE, but such predictions were never used. Even today most Muslims do not accept any predicted date for the new moon to end Ramadan, and yet they happily accept the predicted times for sunsets to end their daily fasts during Ramadan. In fact predicted times are used by most Muslims for the daily prayers (including the start and end times of fasts), but not for the start and end dates of Ramadan. There is obviously a contradiction there.

6.2 Planetary Models

We shall begin here with the early Greek approaches including Ptolemy's model, and then examine the Muslim contributions with the Maragha Model, comparing it with the Copernican approach.

The Early Approaches

From ancient times, the planetary theory assumed a geocentric universe, in which planets, including the stars, the sun and the moon, rotated round the earth in a uniform circular motion. For the convenience of description, we shall use the term planet below as a common name for all the celestial bodies, except when implied otherwise. To start with the original Greek belief was that all the planets move around the earth in perfect circles. This of course contradicted their own observations. In particular the planet mercury was found to be most problematic, sometimes moving even in the backward direction. The Greek scientists tried to find a solution based on perfect circles, since perfect circles represented divine perfection.

In the third and the second century BCE, two slightly different approaches were proposed to account for the difference between the observations and their planetary model. These approaches were based on what were called *epicyclic* and *epicentric* motions, as shown in figure 3 below. In the epicyclic motion each planet moves in a second circle round the earth, while in the epicentric motion the earth is off the centre of planetary motion. Several hundred years later, Ptolemy combined these two approaches into a new model.

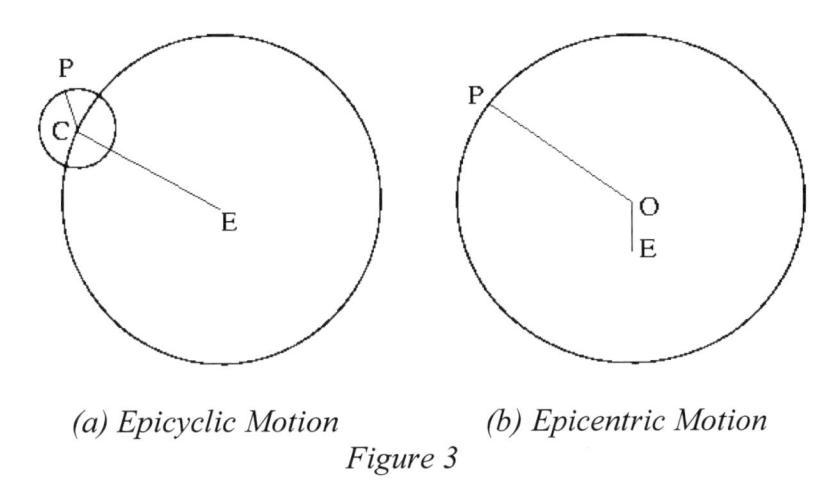

(a) Epicyclic Motion   *(b) Epicentric Motion*
Figure 3

Claudius Ptolemy (85? - 165? CE) of Alexandria was particularly concerned that some stars seemed to be fixed, while others seemed to move backwards. He accepted the epicentric model (figure 3b) for the planets, with two changes. He postulated the centre O of the global circle to be slightly away from the position of the earth E, and the global motion of a planet to be uniform round a different point say F, which is away from both O and E (figure 4). He however, excluded the movements of the moon and the planet mercury, as these could not be explained by such models.

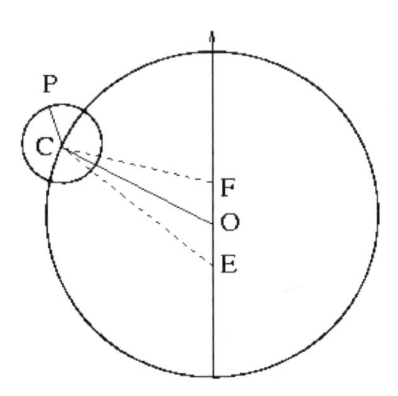

Figure 4: Ptolemy Model

While the Arabics began with Ptolemy's model, they soon noticed its inadequacies, which Ibn Haitham (965-1039) described as "*16 difficulties*". These difficulties included problems with uniform and circular motions. Their attempts, to improve upon and indeed to replace the Ptolemy's model by a better one, began with Ibn Haitham in the 11th century and continued for several centuries, culminating in the work of Ibn Shatir in the 14th century.

Maragha Model

Nasr al-Din al-Tusi (1201-74), survived the destruction of and massacre in Baghdad by the Mongol ruler Halagu Khan, the grandson of Chengis Khan. Tusi, an anti-Abbasid Shia scientist, was able to impress upon and persuade the dreaded Halagu to establish an Observatory at Maragha, Western Iran, in 1265 CE, with he (Tusi) as its Director. Even though Halagu might have been motivated by possible advances in astrology, Tusi was able to gather around him a group of like-minded and bright astronomers for the study of planetary motions.

Already in 1261 CE, Tusi had published his most brilliant work on astronomy under the title: *Al-Tadhkira fi-ilm al-Hay* (Compendium on the Science of Astronomy), which eliminated many of the problems of the Ptolemy's model, as identified by Ibn Haitham. He introduced a more complex model with a device, called the *Tusi's Couple*, which consisted of two further circles, one having half the diameter of the other. The smaller circle moved in the opposite direction at half the speed of the larger one, as depicted in figure 5.

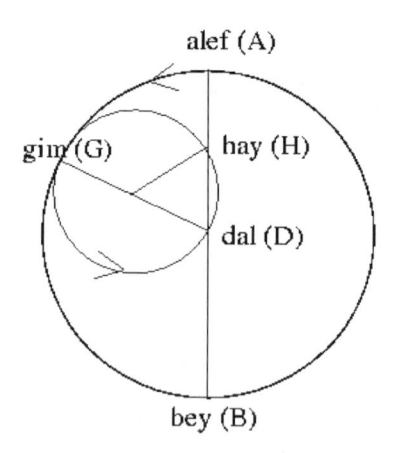

Figure 5: Tusi's Couple

This diagram with Arabic letters (we have given their English/Latin equivalents in brackets) was given in *Tadhkira* and has now become a subject of debate as a possible unacknowledged source of the Copernican model produced some 250 years later (see below).

Besides Tusi, there were also many other prominent astronomers who worked at Maragha, among them: Mu'ayyad al-Din al-Urdi (d 1266) and Qutb al-Din al-Shirazi (d 1311). The latter, who was a pupil of Tusi, continued and extended the work of his teacher.

As regards Urdi, he produced a mathematical theorem for transforming the *eccentric* models into the *epicentric* models. Two mathematical theorems, which later became extremely important in the history of astronomy, were the Tusi's Couple and the Urdi's Theorem, both of which were closely related to the work of Copernicus on the heliocentric model, even though there is no definite proof that Copernicus knew of these ideas.

Comparison with the Copernican Model

Compare the diagram of the Tusi's Couple with the one of Nicolaus Copernicus (figure 6) given in his treatise, *Revolutions of the Heavenly Spheres*, published (in Latin) in 1543 CE on his planetary model. Not only is the presence of the Tusi's Couple obvious (circle GHD nested inside circle ABG), but Copernicus in his Latin treatise has even used the Latin (English) equivalents of the Arabic letters from the Tusi diagram.

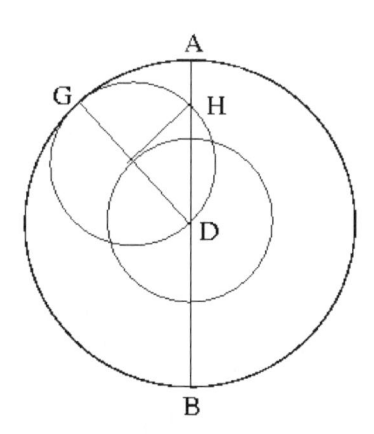

Figure 6: Copernican Model

Yet, Copernicus never acknowledged Tusi's work. Could it all be just coincidence? Historians are very skeptical about such a staggering coincidence. It would seem that a Greek translation of *Tadhkira* was brought to the Vatican some time after the fall of Constantinople to the Muslims in 1453. It is possible that Copernicus, who was Polish, read it there as he was in Italy for a few years and could read Greek, but there is no evidence that he read that translation.

There was another mysterious astronomer called Ibn Shatir (1306-1375), who worked alone in a later age in far away Damascus. He was the time-keeper (*muwaqqit*) of the Umayyad mosque in Damascus. As his job required him to determine prayer times, he delved into astronomy, particularly into planetary motions, as a pastime. He, working in isolation, did the most astonishing thing – he perfected the geocentric Maragha model, which is regarded today as being mathematically equivalent to the Copernican model. But there is a mystery. How was he able to access the Maragha results in an age when there was no transmission and circulation of knowledge outside one's immediate circle and locality? However, it is known that Urdi was Damascene, and his son, also called Urdi, worked in Maragha following his illustrious father; the conjecture is that Urdi junior might have been the link.

The key difference between the geocentric Maragha model of Ibn Shatir and the heliocentric Copernican model was the reversal of the directions of the vectors from the sun to the earth, without changing the rest of the mathematical model. As T. E. Huff stated [Huf95 (p51)]: "...the planetary models of Copernicus, appearing 150 years after the time of Ibn Shatir, are actually duplicates of the models developed by the Maragha astronomers". Huff continues on (p59/60): "The Planetary models of Ibn Shatir and those of Copernicus are virtually identical, with only minor differences in some parameters. But the metaphysical transition would have, of course, forced an intellectual break with the traditional Islamic cosmology, as understood by the religious scholars, 'the ulama'".

However, we may note that the idea of heliocentric planetary systems was suggested earlier by al-Biruni, though not pursued as mentioned above. It was perhaps unknown to Ibn Shatir, or equally it is possible that both al-Biruni and Ibn Shatir did not pursue that line of thought for fear of the orthodoxy. T. E. Huff again (Huf95/p60):

> "The Arabs were perched on the forward edge of one of the greatest intellectual revolutions ever made, but they declined to

make the grand transition from 'the closed world to the infinite universe' ".

The height of Islamic astronomy, indeed, the height of Islamic science was probably achieved with the Maragha model of Ibn Shatir. After this, it spiralled downwards fast. Was it the fear of the orthodoxy? Why did the decline start? These issues will be discussed in Part III of this book. But before we leave astronomy, we shall briefly describe the instruments the Arabic scientists developed for the study of astronomy.

6.3 Astronomical Instruments

There are two sources of our knowledge about Islamic astronomical instruments: (i) instruments that are available in collections (private, public and museums) and (ii) instruments described in treatises. But in neither case there exists any comprehensive inventory. However, some of the most interesting types of instruments that were made and used seem to include :

The Celestial Globe

Ptolemy's Almagest contains the descriptions of celestial globes, armillary spheres and meridian quadrants for the study of astronomy, based on wooden models of Archimedes. Muslims constructed metallic (usually brass) models, hundreds of them, with improvements in precision and with additional facilities. A celestial globe, which was a teaching device, used to contain a model of the sky, with the outlines of constellation figures and other stars, particularly the bright ones, engraved on it in a spherical grid. The problems of spherical astronomy could be illustrated to the students with a three-dimensional celestial globe which included a representation of the sky and the various intersecting planes (e.g. elliptical and equatorial) and angles as discussed earlier. By adjusting the meridian, for instance, the declination and right ascension of a star can be read off from such a globe.

Astrolabes

The astrolabes were the most treasured and sophisticated astronomical instruments in the middle ages, which displayed a mathematical model of the heavenly bodies. An astrolabe can be manipulated to determine, at any time in the year, celestial and time-keeping data, and terrestrial measurements, including prayer times and the qibla. Originally used in the Greek school at Alexandria in the sixth century, the Islamic world learnt about them via Syriac translations after the Arab conquest of Egypt. The Arabic scientists improved and perfected them as very effective and powerful instruments, that could be employed for many astronomical calculations, such as predicting the rising or the setting time of a star (including the sun), and the local time of a place during the day or at night. Mostly made of brass, a basic astrolabe had four parts: a rotatable rete, a grid, pointers representing major stars, and a circle representing the sun's apparent path through the stars. It was first used in the West in October, 1092.

Others

In addition they also built sundials and quadrants. A new instrument called an equatoria was built by the Andalusian astronomers dating from the period (1015-1115 CE), which allowed the positions of the sun, moon and planets to be determined without calculations. Following the observatory built in Baghdad by Khalifa Mamun, a number of other observatories were built throughout the Islamic lands. The one at Maragha was the largest, until another was established at Samarqand in 1420 CE by Sultan Ulugh Beg, himself a mathematician and astronomer. He was a descendent of Halagu and a grandson of Timur (a Muslim). There were other observatories as well, such as those in Afghanistan, Isfahan, Egypt and Andalusia. Numerous scholars were working on astronomy at that time. Most mathematicians also worked in astronomy, from Khwarizmi to Umar Khayyam and Ibn Sina to Tusi, and many more.

References and Sources

[Fak83] Fakhry: "Philosophy and History": The Genius of Arab Civilization, J. R. Hayes (Ed), Second edition MIT Press, 1983.

[Huf95] T. E. Huff: The Rise of Early Modern Science, Chapter 2, Cambridge University Press, pbk edition 1995.

[Ken70] E.S. Kennedy: "The Arabic Heritage in the Exact Sciences", Al-Abhath, Vol 23, 1970, p337.

[Ken73]: E. S. Kennedy: A Commentary Upon Al-Biruni's "*Kitab Tahdid al-Amakin*": An 11th Century Treatise on Mathematical Geography, Beirut 1973.

[Ken75] E.S. Kennedy: "The Exact Sciences", Cambridge History of Iran, Vol 4, ch 10, pp375-395, 1975.

[Ken86] E.S. Kennedy: "The Exact Sciences in Timurid Iran", Cambridge History of Iran (1986), Vol 6, 1986, pp 568-581.

[Hil93] D. R. Hill: Islamic Science and Engineering, Edinburgh University Press, 1993.

[Sab76] A. I. Sabra: Ch 7: "The Scientific Enterprise", Islam in the Arab World, ed. B. Lewis, published by Alfred A. Knopf, New York, 1976, pp181-200.

[Sart27] George Sarton: Introduction to the History of Science, Carnegie Institute of Washington, 1927, comprising three huge volumes (I, II, III), covering the period from Homer to the 14th Century – the size is equivalent to some 5500 A4 pages. Most scientists (including Arabic, Chinese, Indian, Japanese) are individually listed and contributions described. These are indeed the books that made Arabic contributions to science well-known.

[Sin00] Simon Singh: The Code Book, Fourth Estate, 1999.

[Sin96] Charles Singer: A History of Scientific Ideas, Barnes & Nobles, 1996, originally published in 1900 under a slightly different title.

[Tur97] Howard R. Turner: Science in Medieval Islam, University of Texas Press, 1997, ISBN 029278 1490.

CHAPTER 7

OTHER SCIENCES

The Arabic scholars excelled in other sciences as well, including physics (mainly optics), chemistry, geography and mechanics. We shall briefly highlight some of their achievements below.

7.1. Optics

The first Arabic scientist to study optics was Abu Yusuf al-Kindi (d 870 CE, see chapter 10), who provided a precise description of the principle of radiation, following the work of Theon of Alexandria (late 4th century BCE) on rectilinear propagation of light and formation of shadows.

The greatest contribution was made by Abu Ali al-Hasan b. al-Haitham (965-1039), who was born in Basra but died in Egypt. He wrote over 100 treaties, dealing with astronomy, mathematics and optics. The most famous one is entitled: *Kitab al-Manazir* (Book of Optics), which offers a theory of vision. He overturned the earlier mathematical tradition, from Euclid to Kindi, that was based on a certain number of axioms, which were assumed to be self-evident. Instead he developed a complete scientific theory from some basic principles rooted in sense-perception. The principles he arrived at were, in summary:

(i) the visual object must lie at a distance from the eye in a straight line from the eye surface,
(ii) either the light is radiated from the object (if self-lighted) to the eye, or light is reflected or refracted from it to the eye,
(iii) the object must have a size, since very small objects may not be visible depending on the distance.

These ideas were a major scientific breakthrough and provide the basis of modern optics. He carried out experiments to test his ideas, recognising physical, physiological and psychological factors that affect visual perception. Such considerations were all new. He attempted to explain the formation of images in the brain via the optic nerves passing through various layers. He studied radiation, reflection, refraction and the relationship between the angles of reflection and refraction under certain conditions[1]. He also discovered spherical aberration that is seen in the solar eclipses, and wrote on the formation of halos and rainbows. Thus, the basic principles of optics were found first by Ibn Haitham in the 11th century, although Newton was wrongly given credit for this discovery some 600 years later.

However, for some inexplicable reason the work of Ibn Haitham remained unknown in the Islamic world until towards the end of the 13th century, when a Persian scientist Kamal al-Din al-Yusuf came across it. Following Ibn Haitham's work and after carrying out some experiments, Kamal al-Din produced the first scientifically correct explanation of both the primary and secondary rainbows[2], which had eluded even the great Aristotle. The mechanism of inverted-image formation through a pinhole camera (*camera obscura*) was also explained by Kamal al-Din after some experimentation based on the ideas of Ibn Sina (see chapter 10) and Ibn Haitham.

7.2 Chemistry

The Arabic term for chemistry is *al-kimiya*, which also means *alchemy*. In the early research, not much distinction was made between the two. However, so far as chemistry is concerned, some substances were studied, including alcohol, perfumes, acids, alkalis and petroleum. The oil fields at Baku (Azerbaijan) were developed by the Arabics perhaps from around 885 CE. Razi (of medicine fame – chapter 10) in his *Book of Secrets* described the properties of crude oil and its use. The book also explains for the first time how to pro-

---

[1] However, he did not discover what is called Snell's Law that states $\sin\theta = \mu\sin\varphi$ where θ is the angle of incidence to, and φ the angle of refraction from, a material of refractive index μ, thus relating the angles of incidence and refraction through the refractive index.

[2] The actual dispersion mechanism that breaks up white light into the component colours was discovered later by Newton in 1672.

duce caustic soda (sodium hydroxide), an essential ingredient of all detergents, including soap which they invented.

7.3 Geography

There were advances made in geography as well. Khalifa Mamun in particular sent out many expeditions to discover new places. Legend has it that Khalifa Mamun assembled some 70 experts to collate and produce maps. He undertook several scientific projects, including a project to measure the diameter of the earth, under the supervision of astronomer Musa bin Shaker and his three sons. For the diameter of the earth, two groups of astronomers independently measured, with ropes, the distance of one-degree latitude from strict south to north on a flat land, following the pole star. The answer they got was 12,804 kms as against the modern value of 12,744 kms, out by only ½ percent.

Khwarizmi (the algebraist) produced latitudes and longitudes of 545 cities, along with the locations of mountains, islands, seas and other special features. The latitudes and longitudes of many other cities outside the empire were also determined. The surface of the globe was divided into a number of sectors for defining characteristic features. Polymath Abu Raihan al-Biruni (973-1048) introduced the technique of triangulation to measure distances on earth, and one of his treatises, *Al-Qanun al-Masudi*, covers the whole of the medieval knowledge that was available at that time in astronomical geography. Al Idrisi[3] (1099-1166) who is sometimes claimed to be the greatest geographer and cartographer of the middle ages, produced a geographical encyclopaedia with many maps, including a large silver globe (4 Kg) for the Christian King Roger II (b 1095, r 1130-1154) of Sicily.

7.4 Mechanics and Machines

The students of physical sciences were far fewer than those in mathematics, astronomy, alchemy (chemistry) and medicine, and many of them focussed on mechanics. Topics outside mechanics included light, sound and magnetism, but these were not studied in any systematic way. Al-Biruni concluded that the speed of light is much

---

[3] A modern piece of geographical information software bears his name.

greater than that of sound. The physics of sound was explored as part of music, which included an investigation into pitch by Kindi. Both Farabi and Ibn Sina wrote treatises on musical notes, which were far ahead of any Western knowledge at that time. Several of them also carried out some ad hoc experiments on magnetism.

Mechanics was studied in greater detail and many precision instruments (such as the weighing balance) were developed, requiring knowledge of centre of gravity, density, weights, and metals and alloys. The pendulum was invented by Ibn Yunus of Egypt, 400 years before Galileo, for the measurement of time. Another Arabic mastermind, Kutbi, produced the first watch. Other inventions included guns and the ship's navigational compass.

Chinese paper makers were taken as prisoners to Samarqand in 751 CE and were forced to teach this technology to the Arabics. Vizier Yahya Barmakid of Khalifa Harun al-Rashid established a paper mill in Baghdad around 800 CE. It was paper from this mill that permitted so many books to be written and copied. Paper was also exported, and other paper mills established. King Roger II (r 1130-1154) of Sicily used such paper. Europe had its first paper mill in the 12th century, nearly 400 years after the Muslims.

In addition to the astronomical equipment mentioned in chapter 6, they also built water clocks, water wheels, water-raising machines with wheels, gears, axles and clogs, irrigation and water supply machines (and systems), dams and bridges, and mining and surveying equipment. In the middle ages, the Arabics were at the forefront of technological innovation.

References and Sources

[Dal99] Ahmad Dallal: "Science, Medicine and Technology", Chapter 4 (pp155-214) in Oxford History of Islam, ed by J. Esposito, 1999.

[Fak83] Fakhry: "Philosophy and History": The Genius of Arab Civilization, J. R. Hayes (Ed), 2nd edition, MIT Press, 1983.

[Huf95] T. E. Huff: The Rise of Early Modern Science, Chapter 2, Cambridge University Press, pbk edition, 1995.

[Hil93] D. R. Hill: Islamic Science and Engineering, Edinburgh University Press, 1993.

[Ken73]: E. S. Kennedy: A Commentary Upon Al-Biruni's "*Kitab Tahdid al-Amakin*": An 11th Century Treatise on Mathematical Geography, Beirut 1973.

[Sab 76] A. I. Sabra: Ch 7: "The Scientific Enterprise", Islam in the Arab World, ed. B. Lewis, published by Alfred A. Knopf, New York, 1976, pp181-200.

[Sart27] George Sarton: Introduction to the History of Science, Carnegie Institute of Washington, 1927, in three volumes (I, II, III) spanning the period from Homer to the 14th Century – the size is equivalent to some 5500 A4 pages. Most scientists (including Arabic, Chinese, Indian and Japanese) are individually listed and contributions described.

CHAPTER 8

MEDICINE

As stated in chapter 1, the Nestorian scholars established a major centre of learning (called the Academy) and a medical treatment facility in the city of Jundishapur in Khuzistan. In 636 CE the Arabs conquered the area, with its great Academy and the medical treatment centre still intact and fully functioning. Nestorian Jibrail Ibn Bakhtishu was the head of that medical centre during the reign of Abbasid Khalifa Mansur.

It is said that in 765 CE, Khalifa Mansur had a stomach illness, which his court physicians could not cure. So, Jibrail Bakhtishu from Jundishapur was called in and he successfully treated the Khalifa. The recovered Khalifa, marvelling at the medical knowledge of Jundishapur and concerned at its absence among his own physicians, decided that such knowledge should be made available to his own people. He established the first medical school at Baghdad with Jibrail Bakhtishu as its head.

While Jibrail returned to Jundishapur a few years before he died in 771, other members of the Bakhtishu family remained active in the service of the Abbasids. From then on, several generations of the Bakhtishu – all chief physicians at Jundishapur – became chief royal physicians, and thus instrumental in developing and teaching the Galenic medical system (see Galen in chapter 1) in the Khelafa.

The Arabic scholars embarked on this medical enterprise with a great thoroughness, starting with the gathering of all available Greek knowledge on the Galenic system, by translating it into Arabic, sometimes via Syriac. Nowhere was such a project ever undertaken or so successfully completed within such a short time – this was in itself a great cultural achievement in human history.

8.1 The Galenic System

The Arabic physicians used the Galenic system in its entirety without much exception, and developed it further as the best system that was available anywhere for over 700 years from 750 to 1500 CE. In some instances, they criticised the Galenic system for being incorrect, but those criticisms were made after a very careful study. In any case, the orthodox Muslims were suspicious of the idea of secular sciences from the heathens, and particularly disliked the heathen medicine that claimed (as did the Greek medicine) to care for not only the body, but also the soul. It is the care for the soul that was vehemently opposed by the orthodox, who in turn embarked on developing an alternative system called Prophetic medicine (*tibb an-nabi*), sourcing it on a collection of hadiths. Some 80 hadiths (about 2.3 %) of the al-Bukhari collection are on medicine. Later however it was abandoned, at least in the public sector, when it proved to be ineffective. Eventually, the orthodox accepted Greek medicine, though very uneasily.

Galen (129-c199 CE) developed a comprehensive and well-founded system of medicine from his study of animals, in which he showed the importance and functions of various internal organs, such as the spinal cord in muscle activity, the ureter in the kidney, the bladder in urine disposal and arteries for carrying blood (not air as previously believed). However, his idea on blood circulation was wrong. In his system Galen also incorporated general and ethical ideas from the work and practice of Hippocrates of Cos (460-377 BCE), as modified later by Aristotle (384-322 BCE) (see also chapter 1). Hippocrates is the father of medicine, from whom comes the Hippocratic Oath which all doctors today have to take. This Oath requires doctors not to believe any disability or disease to be a divine punishment – they must treat all illnesses for curing without discrimination.

The starting point in the Galenic theory is food, which gets transformed in the body by natural warmth into different substances, some of which are excreted by the body, while others are transformed again in the liver and are transported to the different parts of the body by the blood for useful activities. At the end of this process, four cardinal humours are produced: blood, mucus, yellow bile and black bile. These humours are combined with four qualities: warmth (or heat), cold, moisture (or dampness) and dryness. Examples of combinations are: blood is damp and hot, or mucus is damp and cold. If the four humours with their qualities are in a state of mutual equilibrium, a man is healthy, otherwise the balance is disturbed (unhealthy). A man protects his health by conserving symmetry in different spheres of his life. This is where the point about healing the soul comes in, an idea that was disliked intensely by the Muslim orthodoxy.

The Galenic system was the most advanced medical thinking that was available anywhere else in the world at that time. It was replaced only after the European Renaissance. The Arabic scholars developed the Galenic system further and perfected it. One may contrast the Galenic approach with that of driving out evil spirits, as practiced under Christianity in Europe at that time.

8.2 Healthcare and Practitioners

The practical delivery of medicine and healthcare achieved a high standard in the Islamic empire in terms of hospitals (*Bimaristans*), doctors, carers, medical students, medical schools and books – unparalleled by any contemporary comparison. The first hospital in the world was built in Damascus in 707 CE under the Umayyads, but the first sophisticated hospitals, with all facilities including doctors, wards, nurses, different types of treatments and 24 hour opening, were established in Baghdad under Khalifa Rashid. All subsequent hospitals in the land of Islam followed this one. The concept of modern hospital – as we know it today – has been borrowed from these Arabic innovations.

The physicians needed training and a license to practice. The medical training was thorough. The students were advised how to diagnose and judge, how to relate observations to theories, and how to be precise, brief and clear in describing a disease for patient records. They learnt about anatomy, orthopaedics, fractures (not known in the West until 1852 CE), the use of Plaster of Paris, surgical procedures, anaesthetics, acute infections, and much more besides. There were specialised studies on ophthalmology, surgery, psychotherapy and pharmacy. These were backed up by hands-on-training, with the professors going round the hospital wards with small groups of students. After passing examinations, new graduates were required to undergo postgraduate apprenticeship training and to pass a licensure examination before they were allowed to practice.

All physicians were required to maintain the confidentiality of patients' information and to follow ethical principles based on the Hippocratic Oath. They were forbidden to kill, or help to kill anyone, for instance, by administering a poisonous drug. Pharmacists were employed to check drug preparation and to control quality. Drug adulterations were punished severely. Doctors were restricted from owning and holding stocks in pharmacies. However, some of these ethical controls were allowed to lapse from the time of Khalifa Mutawakkil – that is, roughly after the fall of the Mutazilites. Nevertheless, patient care still remained excellent and improvements continued.

8.3 Hospital Systems

Hospitals had separate male and female wards, along with an out-patient facility that was open 24-hours a day. Hospitals had pharmacies for free distributions of drugs, and special housing for students and in-house staff. Each hospital had a large modern library – one hospital in Baghdad had 80,000 volumes, two in Cairo had 100,000 and 200,000 volumes respectively. The library in Tripoli had 300,000 volumes and one in Cordova had 600,000 volumes. These numbers give some idea of the advanced facilities that were available.

All hospitals (which were Government owned) were required to serve all citizens without any fees and without any regards to colour, religion, gender, age or social status. Records were kept on each patient and their medical care. Poor patients were given some money on discharge. The Qairawan hospital in Tunis (in 830 CE) had female nurses.

Because of its holistic care of the body, the Galenic approach prescribed the consumption of high-quality food as part of the treatment, and as such the cuisine in many of those hospitals was of a high standard – so high that apparently even some healthy people sometimes queued to be patients! There were separate hospitals for lepers, at a time when the lepers were condemned in Europe. Islamic hospitals were the first complex hospitals anywhere, and they naturally provided the models for modern hospital practices throughout the world today.

8.4 Some Medical Giants

Razi (d 925 CE, see chapter 10) wrote descriptions of smallpox and measles, differentiating (for the first time) the two diseases and identifying their symptoms. He also discussed how to dress wounds to avoid infection, recommending alcohol as a disinfectant. Ibn Sina (d 1037 CE, chapter 10) originated the idea of oral anaesthetics, opium being one.

Because of the widescale use of anaesthetics, surgery was more common in the Arabic world, compared to the West at that time. Abu al-Qasim Al-Zahrawi (930-1013) was considered to be the most famous surgeon in the Arabic world. He used cow-bones to make and implant false teeth, while even 700 hundred years later (in the 18th century), US president George Washington wore wooden dentures. They also developed a technique of removing kidney stones by cutting into the urinary bladder.

The Galenic model did however, have a mistaken notion (as mentioned previously) of the blood circulation system in the sense that it assumed blood to seep through minute pores in the wall of the heart separating the two ventricles. It was first corrected by Ala al-din Ali Ibn Nafis (1210-1288) of Damascus from his discovery of the pulmonary blood circulation system, which was later rediscovered in the West by Servetus and Harvey in 1931. The development of the human embryo was described by Ibn al-Quff (1233-1286), a junior colleague of Ibn Nafis.

Razi and Ibn Sina rank among the greatest physicians of all time. Like other Muslim intellectuals of that period, Razi was a polymath. Only half of his nearly 200 written works deal with medicine, the rest are on other topics, such as astronomy, mathematics, philosophy and theology. That he was a great psychologist can be gathered from the title of just one of his works: *On the fact that even skilful physicians cannot heal all diseases and why people prefer quacks and charlatans to skilled physicians*.

One of his most famous books, on *smallpox and measles*, was translated into Latin and then into English and other European languages – it subsequently went through 40 editions between the 15th and 19th centuries! Another of his works *Al-Hawi* (Comprehensive Book) with its 23 volumes, was one of the most extensive medical texts written by a doctor up until the 19th century. It covers almost all medical knowledge known at the time (including Indian and Chinese) and deals with topics such as surgery, clinical medicine, joint and skin diseases, diet and hygiene.

The other towering figure was Ibn Sina. He was a celebrated philosopher and like Razi a polymath, whose writings covered many diverse topics. Apart from being a pioneer of psychology, Ibn Sina is considered to be the greatest medical writer ever. He had described the surgical treatment of cancer, which holds true even today. His master-piece is *Al-Qanun* (the Canon) which covers all Greek and Arabic work in medicine, containing a million words (20,000 pages in today's books!) spread over five volumes. It includes complete studies of physiology, pathology, anatomy, hygiene, fevers, fractures, and even cosmetics, along with discourses on breast cancer, poisons, rabies, amnesia, meningitis, tuberculosis, tumours, kidney diseases and geriatric care.

8.5 Beginning of the End

Until modern times, the works of Razi and Ibn Sina were used as basic texts in most European medical schools, and these were the most used medical references ever. The only modern book that can be compared to this is Henry Gray's Anatomy. However, both Razi and Ibn Sina were declared heretics by the later Muslim orthodoxy (see chapter 10).

Ibn Khatib (1313-1374) of Granada was considered to be one of the greatest intellectuals and writers from Granada. While Christian Europe stood helpless in the face of the ravaging "Black Death" in the mid-fourteenth century, he developed a theory of infection and wrote a book on it. A vizier for some time, a polymath and a friend and teacher of Ibn Khaldun (whose father died of the plague – see also chapter 10), Ibn Khatib was accused of heresy and assassinated in prison while awaiting the verdict of the heresy investigation – this was a sure sign that the decline in medicine had begun.

References and Sources

[Bür76] J. Christopher Bürgel, "Secular and Religious Features of Medieval Arab Medicine", Asian Medical Systems: A Comprehensive Study, edited by Charles Lesley, University of California, Berkeley, 1976, pp44-63.

[Dal99] Ahmad Dallal: "Science, Medicine and Technology", Chapter 4 (pp155-214) in Oxford History of Islam, ed by J. Esposito, 1999.

[Ham83] S. K. Hamarneh: "Life Sciences", The Genius of Arab Civilisation, edited by John R. Hayes, second edition, MIT Press, 1983. pp 173-200.

[Sab76] A. I. Sabra: Ch 7: The Scientific Enterprise, Islam in the Arab World, ed. B. Lewis, published by Alfred A. Knopf, NY, 1976, pp181-200.

[Sart27] George Sarton: Introduction to the History of Science, Carnegie Institute of Washington, 1927, comprising three large volumes, covering the period from Homer to the 14th Century – the size is equivalent to some 5500 A4 pages. Most scientists (including Arabic, Chinese, Indian and Japanese) are individually listed and contributions described. These are indeed the books that made Arabic contributions to science well-known.

[Syed03] I. B. Syed: "Islamic Medicine – 1000 years ahead of its time", Islamic Medicine, edited by S. Athar, 2003, www.islam-usa.com/im4.thml.

PART III

EXAMINATION OF SUCCESS AND DECLINE

CHAPTER 9
CAUSES FOR DECLINE

Picture an evening in ancient Baghdad during the time of Khalifa Mamun. Standing on a dark street one would have seen lights emanating from the large lounges of the noble houses, where lavish parties were in progress – the parties where the great and the good were mingling, discussing and debating excitedly the intellectual challenges of the day. Each participant, like a little candle, would have been lighting up a tiny dot on the winding journey to truth, beauty and knowledge – following the exhortation of the Prophet "Go even to China to seek knowledge". They were the committed pursuers – committed to pursue science, medicine, philosophy, arts and literature. Collectively all these little candles created the power of a beacon, illuminating their path to understand God's mystery, which one day would lead them, they were confident, to the farthest reaches of knowledge. In the process, they hoped, they would be able to create a new civilisation, and this they did. That fever pitch environment was perhaps comparable to the one that existed in the Manhattan Project (the US atomic bomb project) in the 1940's or NASA's moon landing project in the 1960's.

The progress that was made during the Golden Age of Islam was unparalleled – it even propelled the European Renaissance to develop modern science and medicine. And yet, during the past six hundred years, a darkness has gradually descended over Muslim lands everywhere, preventing the cultivation of secular knowledge in all its forms, including scientific knowledge. As the quotation from Professor Salam in the Preface stated:

> "There is no question but today, of all civilisations on this planet, science is weakest in the lands of Islam."

According to T. E. Huff [Huf95/p90]:

"For although the European West was definitely less rich and intellectually sophisticated prior to the twelve and thirteenth centuries, thereafter it underwent a revolutionary transformation, thanks in part to the transmission of a wealth of Greek and Arabic scientific knowledge, which prepared it well for the revolutionary modern science".

We must ask, as did Huff: Why? Was it due to the nature of the social, cultural, legal and institutional practices of the Islamic civilisation that eventually led to the demise of science?

The first nail in the coffin of knowledge was driven in by the conservative Khalifa Mutawakkil, in support of the orthodox, as stated earlier. In an age when everything was dependent on the whim of one individual, viz. the Khalifa, it was easy for him to unleash the wrath of the orthodoxy on what they called knowledge of antiquity, useless knowledge or even as "wisdom mixed with unbelief". The knowledge the conservatives permitted was called useful knowledge which consisted of only religious knowledge from the scripture together with the religious science, a science that was restricted to religious needs, such as finding prayer times.

Gradually over several centuries the yoke of the orthodoxy was allowed to strangle all search for knowledge. They prevented not only the study of philosophy but also the study of universal science as we know it today. Islam became anti-science and anti-intellectual, even though of all the major religions, it is Islam that repeatedly urges its followers to seek knowledge and to understand the mysteries of God's creation.

In this chapter, we wish to examine the causes for the decline of the Islamic science, particularly focusing on those causes that are still keeping the Muslims "weakest" (Salam's term) in science even today. The reasons why the six centuries of Islamic science and intellectual leadership did not directly lead to the emergence of universal modern science, have recently been examined by both P. Hoodbhoy [Hoo91] and T.E. Huff [Huf95].

Huff in particular tried to answer, from the sociological perspective, the question as to "why the Arabic science did not go 'the last mile to modern scientific revolution' " [Huf95/p59]. He has presented a detailed and scholarly analysis, in which he has also compared each issue with the European practices that gave rise to modern science. However as stated earlier, our objective in this book is slightly different. It is not so much as to why medieval Islamic science did not give rise to modern science (although it is an important issue), but rather why it declined and more importantly, what are the reasons that still prevent the Muslims today from developing a community that can thrive in science, as it did in the Golden Age of Islam. Our analysis has led us to the following topics, which nevertheless cover the major points raised by both Hoodbhoy and Huff, but viewed from slightly different perspective:

- Social attitude
- The nature of the state
- Higher education
- Collegiality and dissemination
- Funding support for science
- Inadequate Islamic law

Among these causes, social attitude which includes the impact of Sharia is the most significant one, particularly since its effect is still crippling the Muslim community. It will therefore be discussed in greater length below. The nature of the medieval Islamic state also played a major role in the decline of science, and some aspects of it (such as the rule of law and individual rights) are still highly relevant today. The other four causes were very important in the past, but much less so today; and hence they will be investigated under a single heading of *Other Causes* below. Finally, the issues discussed are highly inter-related, even though, for convenience, we have described them as separate items.

9.1 Social Attitude

Social attitude, so far as science was concerned was formed by the religious orthodoxy, which is recognised by all experts as the principal cause of the decline of science among Muslims. It was the religious orthodoxy that defined both the social and cultural norms in Islam, even though there were cultural differences among various national groups, such as Arabs, non-Arabs, Persians, and so on. The impact of science (or the lack of it) on the society in terms of benefits, also had an inevitable role in shaping social attitudes. We shall address both these aspects below.

9.1.1 Attitude of the Orthodoxy

Modern science assumes the existence of a rational and ordered universe governed by a set of natural laws uninterrupted by any divine agency. It also presupposes man to be capable of understanding these laws, along with the internal structure of the universe, through his own intellect, using rationalism, causality and reason, unaided by any divine guidance. The task of a scientist is to discover these laws and the internal structure of the universe. It is generally believed in physics that there are some deep features, such as truth, symmetry, beauty and elegance, in all these laws and within the structure of the universe. These deep features can inspire men of religion very profoundly.

The early Islamic philosophers and scientists (particularly the Mutazilites) understood this basis of science and therefore placed reason above everything else in laws of nature. This attracted the wrath of powerful Muslim conservatives who could not (and cannot even today) accept that basis. For the conservatives, to believe in any law of nature, one has to believe in a rational universe operated through cause and effect, implying that actions can result from purely those laws without any divine intervention. This to them is the great sin of *shi'rk*, associating the creation of actions to agencies other than God. In fact, the objection to the knowledge of antiquity by the ulama (religious scholars) started long before the first scientist appeared on the Islamic scene. According to the great Egyptian Islamic scholar Abdelhamid I. Sabra [Sab76]:

"From the time when the translation movement began towards the end of the Islamic Middle Ages, these [Greek-based] sciences were either frowned upon or openly attacked by the practitioners of the indigenous religious and Arabic disciplines."

Observe that even the developments in medicine were accepted with great reluctance.

In Islam the idea of freethinking to discover truth and knowledge peaked early in the 9th century, when the rulers were autocratic, far removed from the masses. Intellectual activities were confined to the higher strata of the society, solely dependent on the conviction of one person, namely the Khalifa at the top of the pyramid. Even when a Khalifa supported the cause of freethinking, he did not mind persecuting those who did not agree with his version of freethinking. Khalifa Mamun instituted a *mihna* (minor inquisition) against those who were opposed to his Mutazilite doctrine, and Khalifa Mutawakkil used the same *mihna* to persecute those that supported the Mutazilite philosophy. Thus, unlike the reformation in Europe, the freethinkers in Islam lost out to the orthodox in the middle of the 9th century, when the downfall began. Even if the freethinkers had not lost to the orthodox, it is unclear whether an Islamic renaissance would have taken place, since a renaissance would have required evolution into more human rights and more representative governments. Given the subsequent history, such an evolution is unlikely to have been permitted by successive later rulers.

In the early days of the ban on freethinking, the religious zealots appeared in the form of the Hanbalites and the Asharites. In 885 CE, they even barred the professional copyists in Baghdad – the printing press of the day – from copying the books of the philosophers. They first attacked the Mutazilites, and then the other philosophers, and finally the whole of science (including logic and mathematics). Later, the scope of attack was extended to include anyone who wrote anything that could conflict with the orthodoxy. Even Ibn Khaldun who was an anti-Mutazilite was denounced as a heretic for his *asabiya* theory (see chapter 10). The denunciation was initiated by Imam Ahmad Ibn Hanbal who aroused the populace with "fire and brimstone" against the philosophers, scientists, and the promoters of reasons, with the backing of the new Khalifa Mutawakkil, among the cheers of the masses. Anti-Mutazilite riots took place and their books and libraries were burnt. He declared "Every discussion about a thing that the Prophet did not discuss is an error".

Ponder for a moment the consequence of such a doctrine of closure. No philosophical, scientific, technological or cultural progress would ever be possible – no paper, no printing, no electricity, to name but a few items. Recall, Imam Hanbali even refused to eat watermelon, since he could not find any precedence that the Prophet ever ate it. We shall present below a few more examples of the reactions of the orthodox to bring home the extent of the anti-science environment that was created.

Imam Hanbali was so successful that his own grandson Abd al-Sallam, who had an interest in philosophy and magic, had his house searched, books burnt, and was finally condemned as a heretic for keeping/reading forbidden books. The celebrated orthodox sufi Abdul Qadir Gilani was a family friend. Both Imam Hanbali and Abdul Qadir Gilani were declared guilty as *possible* teachers, and an official curse was pronounced on them as punishment! Although a curse today is laughable, the incident nevertheless demonstrates the existence of an anti-intellectual atmosphere in which not only the philosophers, but also their friends and relatives, however revered, were declared guilty by association. As for Abd al-Sallam himself, he was dismissed from his academic post, imprisoned and later released when he confessed his sins.

The strong intellectual opposition to the freethinkers was launched by Imam Gazzali, who denounced the *falsafa* (the Islamic philosophy founded on Greek thought) practitioners individually as heretics (see chapter 10). According to Fazlur Rahman, a highly-respected Pakistani Muslim scholar[1] [Rah84/p133]:

> "… al-Gazzali drew the fatal conclusion that therefore people should be discouraged from studying even the scientific works of the philosophers".

Akbar Ahmad, another eminent scholar, originally from Pakistan, concurred [Ahm98/p85]:

> "Although it was not his primary intention, Gazzali had effectively sealed the fate of Greek thought in Islam. Henceforth it would be seen as bordering on heresy. This reaction easily extended to other non-Islamic philosophy. In the realm of ideas it became the second nature for Muslims to reject non-Islamic philosophies. Muslims would be encouraged to be inward-looking and self-sufficient. They would be quick to dismiss original or fresh thinking as *ijtihad* or innovation and to condemn it."

Over the years an inward-looking Islamic self-image of superiority and self-sufficiency was developed, which treated any idea imported

---

[1] See chapter 13 footnote 8 on Fazlur Rahman's contribution to Islamic thought.

from outside as *bida'* (see also later). This permitted only the discovery of existing knowledge, but not the creation of new knowledge by learning from other cultures in order to meet new challenges.

As regards Gazzali, he was equally clear on science:

"It is rare that someone becomes absorbed in this [foreign] science without renouncing religion and letting go the reins of piety within him".

"Even if geometry and mathematics, do not contain notions that are harmful to religious belief, we nevertheless fear that one might be attracted through them to doctrines that are dangerous".

The conservatives particularly objected to figures in geometry. The owner of a book containing mathematical representation was condemned as a heretic. Geometry was often seen as a threat to faith. Ibn Haitham's (965-1039) book on astronomy was also treated with horror for the geometric representation of the planetary movements. An orthodox *alim* (religious scholar, singular of ulama) complained of this geometric figure for representing "shameful temptation, speechless calamity and blind misfortune". The abstract nature of mathematical thinking was also considered offensive as it "weaken[ed] the faith". Although Gazzali himself was not opposed to logic (in fact he used it), the other ulama and the masses believed that "He who practices logic, becomes a heretic". After the death of Gazzali, the opposition to the study of logic gathered pace and became more strident.

Ibn Salah al-Shahrazuri (d 1251) studied logic under the great polymath Kamal al-Din al-Yusuf who explained the rainbow correctly (chapter 7) and also taught Nasr al-Tusi of Maragha. But Ibn Salah failed to achieve a pass certificate in logic, as he could not satisfy his teacher by his performance. When the same Ibn Salah became a great hadith expert, he issued a fatwa in which he called all those who taught logic and philosophy (which included his former teacher Kamal al-Din) either to be executed or converted back to Islam. His fatwa continued [Gol81]:

"It is the duty of the civil authorities to protect Muslims against the evil [i.e. logic and philosophy] that such people can cause.

The persons of this sort must be removed from the schools and punished for their cultivation of these fields. All those who give evidence of pursuing the teachings of philosophy must be confronted with the following alternatives: either (execution) by the sword, or (conversion) to Islam, so that the land may be protected and the traces of those people and their sciences may be eradicated. ... . The eradication of evil must involve the eradication of its roots. The employment of such a person as a teacher in a public school is a most abominable thing".

Ibn Salah did not spare Imam Gazzali either for applying logic. What Ibn Salah did was to provide a formal expression, in the form of a fatwa[2], of an existing practice. The persecution of Saif al-Din (d 1234 CE) was a case in point. He was a scholar of the Shafi'i and Hanbali schools, but studied the forbidden logic and philosophy as well. In Cairo, he was persecuted by the ulama for his study of philosophy, particularly logic, even though he did not teach either of them. He was accused of having perverted his faith, and a fatwa was issued condemning him to death. He fled to Damascus, where he managed to get a job, but later he lost that too, due to a similar allegation.

A similar hard attitude towards philosophy was taken by Taj al Din Subki (d 1271) and Hanbalite Ahmad Taqi al-Din Ibn Taymiyya (1263-1328) of Damascus. Taj-al Din Subki, a renowned Shafi'ite scholar, denounced and even attacked the later proponents of *kalam* for incorporating philosophical tracts. It would seem that in attempting to ban all philosophy, he claimed (with justification) to have allied himself "with the great majority of the Imams and sheikhs, and the sheikhs of our sheikhs" [Gol81]. It became customary to denounce the intellectual activities with the invocation: "May God protect us from useless knowledge".

According to Subki, some views held by Gazzali, such as on logic, were unacceptable to the contemporary orthodox, who (and the general public) blamed science (which Gazzali studied) as the cause of his deviations, implying that science should not be studied.

---

[2] A fatwa (a legal opinion) can be issued by any Muslim scholar, and therefore it is always possible to find a favourable scholar to issue a desired fatwa. Once someone is denounced by a fatwa as a heretic, no amount of counter fatwas can remove entirely the suspicion from the minds of the masses.

The orthodox demanded scholars to avoid association with persons involved in the science of antiquity, and to be aware of the danger such persons represented. This demand became particularly vigorous after some well-documented incidents. There was one Abu Mashar al-Balkhi (d 885), who satisfied the orthodox as a pious scholar, particularly after he reportedly incited the people against the philosophy of Kindi. But subsequently on his way to the hajj from Khurasan, he visited the rich library of the vizier Ali bin al-Munajjam, where he became so engrossed in astronomy that according to the orthodox, he forgot the hajj and his religion, thus becoming "a heretic". What Abu Mashar did later, we do not know. Perhaps he deferred the hajj (which is permissible), and/or he became less orthodox, but either would probably have attracted denunciation from the orthodox, particularly since the study of a forbidden subject (that is, astronomy) was involved. Following many such incidents, some scholars, for example, Laith bin al-Muzaffar (editor of Kitab al-Ain) refused to study astronomy, for fear of going astray. Thus an anti-science climate was established. Not only was there no incentive to practice science, but also there was a real danger for the practitioners and their friends and relatives.

Hanbali Imam Taymiyya of Damascus (1263-1328), mentioned earlier, was stricter than Imam Hanbal himself and he became the forerunner of the Wahhabi sect of Saudi Arabia. He viewed all knowledge that was not derived from the Prophet to be useless knowledge. Avoiding contradiction with the hadith that urged Muslims to travel even to China to seek knowledge, he declared '*ilm*' (knowledge) to be only religious knowledge, one that was derived from the Prophet. Clearly this would imply that learning papermaking and printing from the Chinese, or mathematics from the Indians, not to speak of medicine from the Greeks, were all wrong. He even chastised Imam Fakhr al-Din al-Razi (1150-1209), not to be confused with *Razi* of Medicine fame in chapter 10, for writing a book on astronomy. Ibn Taymiyya's anti-falsafa view was recommended to his followers by his disciple Shihab al-Din al-Haytami who forbade the study of astronomy, "as it leads to things that are detrimental", such as "the belief that the world has no beginning". It should not come as a surprise to learn that the Saudi Wahhabi society regards the Islamic Golden Age of the early Abbasids as profoundly un-Islamic!

According to Ignaz Goldziher [Gol81], the Hungarian Islamic scholar, two devastating charges brought by the orthodox were:

(i) that the study of logic, ancient science and philosophy made one disrespectful of religious laws, and
(ii) that such knowledge was useless and ungodly, except where the study was restricted to purely religious matters.

Even the study of geography became useless knowledge, and hence abandoned, with the result that Khwarizm-shah, Abul Abbas al-Mamun (1009-17), decided to execute a traveller as a heretic, for telling him of a country where the sun shone at midnight. Al-Biruni (973-1048) who was at the court at that time was able to stop the execution by authenticating the tale[3]. There is always a price to pay for the cultivation of ignorance. The Ottoman knowledge of European geography was so poor that they did not even know that a ship could sail from the Baltic to the Aegean Sea through the Strait of Gibraltar and across the Mediterranean Sea. When suddenly in 1770 the Russian fleet (see chapter 11) confronted the Ottomans in the Aegean, Turkey lodged a formal complaint to Vienna for letting the Russians sail to the Adriatic (through a non-existent canal) by the side of Vienna! Compare this to the Golden Age of the Abbasid rule, when it was the Muslims who were the masters of geography, amongst other sciences.

Rational interpretation, let alone critical interpretation, of scripture was also forbidden, as it is even today. In 1930, Mohamed Abu Zaid, an Egyptian Sheikh, published a translation of the Quran, in which he criticised the old commentaries and gave simple naturalistic interpretations to the supernatural references. It was promptly confiscated by the authorities and the translator was banned from

---

[3] Khwarizm: A sultanate that developed during the fragmentation of the Abbasid power. It covered roughly the area of present-day Uzbekistan and Turkmenistan, and by the end of the 12th century, even Khurasan (a region in modern Iran) and Transoxania (the present-day Kazakhstan and part of Uzbekistan).

It is possible, but unlikely, that the said event took place towards the end of the rule of Ali al-Mamun (d.1009 CE), the elder brother of Abul Abbas, with al-Biruni at the court. However, returning to the argument, if the sun shone at midnight, then it would make a nonsense of the prayer and fasting times, and therefore it could not be true – thus went the argument before al-Biruni intervened.

preaching his ideas. More recently the Saudi Government set up a committee to produce an English translation of the Quran. After sitting for several years, it gave up the idea of a new translation, but decided instead to approve the scholarly translation of Yusuf Ali, while removing some of his more rational interpretations and replacing some of them by those that are acceptable to the Wahhabi orthodoxy. This version is now one of the two officially approved Saudi translations of the Quran in English.

Impact of *Bida'*

Another objection to the study of science and technology was the concept of *bida'* in Islam. It is assumed to mean innovation (but later interpreted to imply anything outside custom and practices legitimised by Sharia), for which the following hadith (of dubious authenticity) is often quoted by the orthodox:

> "The worst things are the novelties, every novelty is an innovation, every innovation is an error, every error leads to hell-fire".

In its extreme it advocates the rejection of all new ideas and amenities that were not known to, commented upon or enjoyed by the Prophet; which is every thing that the modern civilisation offers – from electricity to computers, including modern medicine and even tap water and nappies. However, even for Muslims in the middle ages, this hadith became too hard to follow, such that the arch-conservative Imam Ibn Taymiyya had to concede that this total *bida'* was unworkable. So he developed the notion of good and bad *bida'*s, only the bad ones to be forbidden – the bad *bida'*s were the ones that were contrary to the Quran, hadiths and *ijma'*. However, since *ijma'* depends on the climate of opinion which may be based on needs and may vary from place to place and time to time, today's *bida'* can become the tomorrow's sunna (tradition).

To take some examples, the study of English or the use of radio, telephones, etc were *bida'* before, but are not now. For instance microphones (blasphemously imitating the God's gift of human voice) are no longer *bida'*, being now routinely used in mosques including in Mecca. The Lucknow[4] clerics who declared the use of the Internet as haram (as a *bida'*) in 1996, have now seemed to have accepted the Internet as a valuable facility. But in the meantime they have given every discouragement to Indian Muslims seeking to learn science. They are now falling further and further behind in comparison with their non-Muslim compatriots. It's a tragedy how this term *bida'* has been used liberally to make Islam illiberal, and to make it anti-intellectual, anti-science and anti-technology.

According to Fazlur Rahman [Rah79/p107]:

"Theology monopolised the whole field of metaphysics and would not allow pure *thought* any claim to investigate rationally the nature of the universe and the nature of man".

The rational philosophy in the Hellenistic tradition came to be known in Islam as *ilm-al Awa'il* (knowledge of antiquity), which was eventually equated with heresy by the orthodox tradition. Thus ended the era of rationalism in Islam, replaced by that of orthodoxy. Whenever rationalism is abandoned, the door invariably opens up to fanaticism and extremism, as has happened in orthodox Islam.

There is another point about the Muslim psyche that should not be forgotten. The successes in the earlier periods were interpreted by the Muslims as God's blessing for being good Muslims. To earn even more blessings, the ulama urged the society to be even more religious, which actually meant more orthodoxy. The trauma of 1258 (Mongol invasion of Baghdad, section 2.4) – which was far worse than that of the September 11 to the US – was interpreted by the ulama as a sign of God's displeasure for religious laxity of the Muslims, and therefore the return to lost glory was assumed to lie in even more orthodoxy. Every time the next set of rulers, particularly the Ottomans, promised stricter adherence to Sharia. The "un-Islamic activities" such as foreign science and philosophy had to be stopped. A symbiotic relationship was created, the rulers supporting the ulama and the ulama supporting the rulers, as in Saudi Arabia today. This is why reformation in Islam is so hard.

---

[4]Actually Deoband near Lucknow. This centre has been the foremost (and orthodox) Muslim centre of religious learning in the Indian Subcontinent for a few centuries. It lives in a world of its own, unadulterated by reality. It is this centre which in the 19th century declared the study of English as haram, a fatwa that helped keep the Indian Muslims backward.

9.1.2 Social Impact

Science was supported by some Khalifas, but not by the masses, who did not share the enthusiasm of the scientists, nor did they understand the scientists. The intellectuals themselves treated the masses with contempt, regarding them unfit for higher thought[5]. Teaching of those higher ideas were restricted to the scholars in the princely houses. On the other hand the ulama were always dependent on the masses, who never understood the intellectuals and their ideas. They preferred the ulama with their simple black and white truths, threatening with the wrath of God and hellfire for deviations from the strict orthodoxy. Therefore it was relatively easy for the ulama to denounce science successfully.

However, had science permeated through the social ladders and got woven into the social fabric and if the masses saw and derived benefits from the scientific endeavours, then the outcome could have been very different. One exception was medicine, which despite the initial objections of the orthodox, was flourishing so much that they had to accept the Galenic medicine in preference to the so-called Prophetic medicine, *albeit* reluctantly. In this section we wish to explore the impact of science (excluding medicine) on the medieval Muslim society.

Algebra invented by Khwarizmi was quite elementary, it allowed a process of solving quadratic equations (as illustrated in chapter 5), and it was improved further for higher powers by Kharaji and Umar Khayyam. But it never took root in the Muslim mind, nor was it used for practical problem solving – which explains why it was later treated as useless knowledge, so much so that even its study was forbidden.

Although the Indian decimal numbering system was introduced by Khwarizmi in the 9th century, it was never practiced in any coherent way under Islam. In fact there were three types of numerals that continued to be used by Arabic people: (i) arithmetic of the scribes in which the numbers were written in words, (ii) Indian decimal numerals and (iii) the *abjd* system for the astronomers (see chapter 5). It is surprising that the astronomers who had to carry out

---

[5] The idea of popularising science or philosophy to the masses is a relatively recent innovation, even Newton was contemptuous of the masses.

complicated calculations using the tedious sexadecimal numbers failed to see the advantages of decimal numbers.

As stated earlier all scientists worked as individuals, there was no scientific society or establishments to harness and advance common facilities. This demonstrated a major failure of the Arabic scientific community to create a common numbering system, even though the Indian system was available. It was left to the Europeans of the 16th and later centuries. Again the reason was individualism in scholarship, without any community or state backing for innovations. There were groups, in which the group members drew from one another, such as the Maragha group of astronomers, but the information flow never permeated group boundaries, and in this sense it was essentially individualistic. Because of this blackout in transmission, the question as to how Ibn Shatir was able to get access to the Maragha results became a mystery, as discussed in chapter 6.

As stated earlier, after learning paper-making from the Chinese, the Muslims started large scale paper production around 800 CE. Fortunately the paper technology arrived during the Mutazilite period when all innovations were encouraged. But in later years a curtain of darkness descended over the Islamic world, as the orthodoxy declared all forms of new knowledge and innovation as *bida'* and hence haram (forbidden). Anti-innovation was the prevailing climate when the printing technology was brought from China to Islam in the 13th century. The ulama promptly refused to permit its use for the "holy" languages (which were Arabic, Persian and later Turkic), preventing the spread of knowledge among Muslims, as discussed further in section 11.2.

While much scientific research was carried out on selected aspects of astronomy, it was never applied in the critical areas of predicting the new moon for Ramadan and *Eid* festivals, as discussed in chapter 6. Even today, when the Americans can land a man on the moon, the Muslim society[6] does not in general believe calculations for religious events – physical sighting is the only one that counts. As a result, *Eids* are celebrated often on different dates even in the same country. Since the Prophet is not known to have used calculations for the new moon, the orthodox would not commit a *bida'* by permitting calculation-based prediction!

---

[6] In some minority Muslim sects, such as the Ahmadiya, Eids dates are fixed by calculations.

There were however some applications of science with technological innovation, such as in the building of bridges, dams, hydraulics and above all medicine. But non-medical applications were too few and not widely visible for the populace to appreciate the benefit of science. For them it was a godless subject, best to shun, rather than take a risk of hell-fire in the after-life.

No matter how great was the progress the Muslims made in their sciences, these had no roots in the fundamental aspirations of their governments, establishments and communities. Researchers became so isolated that they had nothing to give to their community, which the ordinary people could accept as essential enrichment of their lives (Sau63/footnote).

9.2 The Nature of the State

We wish to examine here the nature of medieval Islamic state, the power and authority of the rulers, the involvement of the state to its economic progress that dictated the technological development, and the rule of law providing protection of those engaged in intellectual and scientific activities.

9.2.1 The Rulers and the Ruled

The Islamic state was not based on the principle of representative government as we understand it today. The Prophet did not select his successor, but this *sunna* of *not* selecting the successor was abandoned by the first Khalifa Abu Bakr in order to prevent potential chaos. Following him, Khalifa Umar set up a committee to select his successor. Muawiya became the first Umayyad Khalifa by force, and then he selected his son Yazid as his successor, claiming that he was following the *sunna* of Abu Bakr and Umar. From then on the use of force became acceptable as a legitimate means of succession, resulting in many sultanates in the lands of Islam. Increasingly these rulers entered into understandings with the ulama, who declared them the legitimate rulers in the Friday Khutbas in return for more Islamic orthodoxy. Even the tyrannies of these rulers were tolerated on the principle of Imam Ibn Taymiyya that "sixty years of an unjust imam is better than one night of anarchy" [Ruth91/p179]. Even the execution by a ruler of all his potential opponents, especially of his brothers – even baby brothers as practiced by the Ottomans – became acceptable, in spite of the emphatic Quranic injunction forbidding the killing of any innocent human being. Sharia did not prevent such killing, in fact the Turkish ulama often approved executions of the contenders with official fatwas, ostensibly to avoid chaos (see chapter 11). Sharia does not cater for human rights.

As it happened the ruler and the ruling elite never felt entirely bound by Sharia in their personal conduct. For instance, the rulers and the princes could (and did) kill anyone they wanted without any accountability, and sometimes even innocent people for fun. When a man claiming to be very strong was presented to emperor Jahangir (chapter 11), he decided to test the man's strength by forcing him to fight unarmed a hungry tiger. The emperor enjoyed the gory spectacle and when the man was torn into pieces by the tiger, he commented that the man was not that strong after all. Another example is when the army went to kill Ibn Sina quite openly, because they did not like his religious views (chapter 10). No subject had any right, not even the right to life. At best they could only have privileges at the pleasure of the ruler. There was no Magna Carta in any Islamic state guaranteeing individual rights.

9.2.2 Economic Disconnection

The main sources of income in the Islamic empires were trade, manufacturing and agriculture. The major manufactured and primary products comprised paper, textiles, carpets, shoes and some minerals, but agriculture was the primary source of income. The Quran and hadiths acknowledged the importance of business and trade, without any comparable comment on agriculture, presumably because it was not central to the Arab way of life at that time. But when agriculture did become central, the age-old Arab aristocratic tradition of ignoring it still continued and in fact became the norm for the elite. An undesirable side-effect of this attitude was that the subsequent Muslim rulers paid very little attention to agriculture, even though it was their main source of income. The result was an economic disconnection of the Muslim elite from the agricultural producers.

Khalifa Umar declared the agricultural lands in the conquered territories as state property (see chapter 4). Subsequent rulers treated these agricultural lands largely as their personal property, which were distributed to others initially permanently, but later temporarily, purely for tax collection from the peasants tilling them. These collectors would typically levy a tax and contract other tax-collectors, who would in turn levy a higher tax, the difference being their own income. There was a hierarchy of such collections, each with higher taxes, the peasant paying up to half of their produce as tax. In Turkey some of the tax-collectors were military generals, who were given agricultural lands in lieu of salaries for the duration of their services. The elite, property owners and the collectors had no interest in agriculture, apart from tax collection from the peasants in order to maintain their extravagant life-styles.

Islamic society was largely urban, with open contempt for the villages, the farmers and the peasants. In fact no distinction was made between a farmer and a peasant, both were equally despised[7]. As long as agricultural production was in surplus and the tax income was coming from the peasants, the elite and the town-dwellers did not care. Since it was beneath the dignity of a respectable man to be a farmer or a peasant (the latter always lived at a starvation level) and since all lands belong to the ruler, there was no evolution of landed gentry in Islam, a gentry that could have formed capital for investment and for eventual technological innovation, as happened in Europe. Apart from some small-scale irrigation projects in some selected areas, agriculture remained untouched by any improvement or by any labour-saving devices[8].

The industries were mostly small-scale, with abundance of labour but hardly any funds to invest in any mechanisation. The elite were interested mainly in the final products, with little thought for the producers, their needs or the future of these industries. There were a handful of large industries owned directly by the elite, but even there they failed to innovate. For instance the Muslims had a monopoly of paper-making, supplying Europe from 1150 CE, but they lost this advantage by the year 1300 CE when France started the export of high quality paper to Egypt. Before long the unprotected Islamic industries were decimated by the cheaper imports from the mechanised West. There was one manufacturing sector in which the

---

[7] The remnant of this attitude prevails even today in the use of the Arabic word *madani* (city-dweller) as a term for respect and *fallahin* (strictly peasants, but often the villagers) as a term for contempt.

[8] Interestingly, this also explains why in spite of a flourishing slave trade, slave-labour (except as domestics) was not common in Muslim countries – there was no incentive to reduce labour cost in production systems [Sea03].

state was particularly keen, it was weapon production, which was again based on the existing technology with little innovation. In contrast the West was investing heavily in science and new technologies which included weapon production and innovation. When new weapons from Europe arrived in the lands of Islam (Mughal India, Iran, Egypt and Turkey), it was too late for the Muslims to catch up (see chapters 11 and 12).

9.2.3 Rule of Law and Protection of the Scholars

The ruler was above the rule of law, as were his friends and relatives. If the ruler gave protection to someone, that person also stood above the rule of law, including Sharia law. Equally if he wanted to harm a person, no law including Sharia, could have protected that person. In fact there was no rule of law, no accountability and no individual rights. Everyone in those regimes lived in fear of the ruler, who could confiscate property, imprison a person or even execute him, at the slightest misgiving. Khalifa Abd al-Malik killed a courtier for *reportedly* making a joke to some other person about the well-known bad breath of the Commander of the Faithful, and Mansur killed his loyal friend for giving advice he did not like, although the advice later turned out to have been sound. Khalifa Rashid killed his able vizier for no good reason. Jahangir (before he became emperor) killed scholar Abul Fazl, an advisor to and a close friend of his father Akbar the Great, as he did not like Abul Fazl's influence on his father. In the Mughal court, whenever the emperor uttered a word however trivial, the courtiers would hail it (and were expected to hail it) with heads bowed, as a pearl of wisdom dropping from the lips of the omniscient and omnipotent *shadow of God on earth*. It was not wise to hold a view, even an intellectual one, that contradicted the wisdom from that omniscient one.

Since Sharia held an anti-intellectual and anti-science position, the only way an intellectual or a scientist could survive was by the patronage of the ruler. Was this a dependable basis for science to flourish? Kindi and Razi were disgraced, Ibn Sina escaped death narrowly, Ibn Rushd was banished and Suhrawardy was executed (chapter 10). The pioneering scholarship of Kindi, the medical genius of Ibn Sina and Razi, or the even chief qadi-ship of Ibn Rushd could not shield them from attack. If these giants of Islam were treated like this, how could the others, the ordinary mortals, dare to take up a life in science? They were frightened and did not dare.

Dependency on the rulers became more precarious as time went on, with the rulers becoming increasingly more dependent on the ulama. New rules arriving at the scene required the continuous support of the ulama for legitimacy through the Friday Khutbas. The inevitable concordats between the rulers and the ulama did not exactly encourage scientific and intellectual activities to flourish.

It may be stated in passing that the separation of religious jurisdiction from the secular jurisdiction of the rulers was an important precondition for the development of Western science as an autonomous enterprise. The traditional Islamic world-view does not permit this. An Egyptian scholar Ali Abd al-Raziq (1888-1966) a judge in Alexandria and a lecturer in a madrasa there, wrote a book entitled: *al-Islam wa usul al-hukm* (Islam and the Bases of Power) in 1924, after Turkey had abolished the Khelafa. In this book he argued against the need for a Khalifa, on grounds that the power of the Prophet was essentially religious and spiritual, not political. A court with 24 ulama from al-Azhar passed a judgement of heresy against him, disqualifying him both as a judge and as a lecturer. He had to leave Egypt for Paris where he got a job. His crime was that he advocated the separation of religion from politics.

Even today in the 21st century, the Muslim states do not have democratic institutions nor any accountability, nor any regards for human rights, even though some of them claim to be democracies – a point to which we shall return in chapter 14.

9.3 Other Causes

As explained earlier, we shall cover in this subsection the other four causes that led to the decline of science under Islam. These causes are higher education, collegiality and information dissemination, funding support for science and the inadequacy of Islamic law in protecting a scientist (as an individual) or his invention.

9.3.1 Higher Education

For science to flourish, it needs high-quality centres of higher learning which encourage freethinking and questioning for the exploration of new ideas. A reason why Britain produces so many original ideas in both science and other spheres of life is its education system and social attitude. Both of these encourage diversity in ideas so that every child can develop its own individuality with its differences generally appreciated, rather than routinely pilloried. Under medieval Islam there were no institutions of higher learning, no universities in the modern sense and no research centres for the new scholars to develop their potential. In the Abbasid period, higher learning was dispensed through the parlours of the Khalifa, the princes and other patrons of scholarships, who also provided the financial support. There were some public libraries, but many of these patrons and scholars also provided private libraries for the use of students. However, this was essentially an ad hoc individualistic system, that fell apart when the Khalifa withdrew his patronage.

The madrasas with emphasis on hierarchical ideas and memorising were the only institutions of learning available. Most madrasas were funded by charitable private trusts (*waqfs*) of single individuals, which could and usually did, stipulate who (for example, from which family) could be appointed to which positions. As religious endowments, these madrasas were required to follow not only the wishes of their founders, but also the religious orthodoxy of the day. These were established within a school of Islam (i.e. the madhab), the curriculum typically being the study of the Quran, hadiths, religious doctrine and Sharia (different for different madhabs). Arithmetic was limited to handling fractions to calculate the share of inheritance. Philosophy, medicine, algebra, higher mathematics, astronomy physics, optics and chemistry (that is, all the known branches of science at that time) were formally excluded from the syllabus. Subsequently some higher madrasas permitted a limited admission of astronomy, logic, mathematics (to cater strictly for the religious needs) and some medicine.

There was no system of examinations as we know it today. The teacher would decide if a student had mastered all his (teacher's) subjects by reading, copying and memorising the manuscripts, then the teacher, if satisfied, would issue an *ijaza* (license) to the student to teach these subjects to others. It was modelled on hadith transmission with the names of the previous successive teachers as the *isnad*. There was no other authority, save that of the teacher of the subject, to validate the competence of the student or to question his knowledge. There were no external validations of the standard of the courses, of the teachers or of the examinations. The madrasa itself, the state or even the Khalifa had no say in what a teacher taught or on the standard a student had attained.

Teaching was highly personalised. In some madrasas, some teachers were great scholars, but their ideas were taught in isolation, without any interaction with the ideas of other scholars. Once a scholar retired, he would select someone (normally his son) to succeed him in that post. Thus there was no input of fresh ideas or knowledge into the system. Some bright students, such as young Gazzali, would take lessons from several masters, get *ijazas* from each, and then combine these teachings in his own mind to create fresh ideas. But this was very rare. Obviously the syllabi were strictly controlled, permitting only the religiously approved subjects. There were no formal structures of teaching and assessment, nor any control, nor any interactions of alternative thoughts to help arrive at a critical understanding. Each teacher was completely independent. Teachers were held in great esteem, questioning a teacher on anything was considered irreverent. It was worse in religious matters, in which the old masters were given an exaggerated veneration.

There was no possibility for science and philosophy to spread through these madrasas, which were designed to glorify existing religious knowledge, but not to create new knowledge by learning from all diverse cultural sources in order to meet new challenges. Even in the most celebrated centre of Islamic learning, Al-Azhar (at Cairo), no professors read any book in any European language until about 1950. How can they have a modern world-view?

9.3.2 Collegiality and Dissemination

In the Western countries, there were formal institutional frameworks for the development, improvement, dissemination and application of knowledge, from which their universities and scientific societies (or guilds) evolved. There was no comparable framework in the Islamic countries. Exceptions are the schools (madhabs) of Islamic law, which can be seen as a kind of guild, defending, cherishing and propagating the ideals of the respective madhabs. In contrast, there were no equivalent societies or institutions to defend, improve and propagate sciences or to fight for scientists. It is often argued that the personalised nature of knowledge discouraged the formation of such scientific societies. However, during the Mutazilite period, the Baitul Hikmah and the Khalifa himself provided the necessary support and encouragement. But later this support disappeared, making scientists vulnerable to attack from the orthodox, to such an extent that even al-Biruni felt obliged to join the orthodox in attacking Dr Razi (see chapter 10). It was not too difficult to persuade some orthodox ulama to issue a fatwa against someone – in fact there were an abundance of ulama ready to denounce scientists, and thereby to score points.

The brilliant work of Ibn Shatir on planetary motion was never followed up by any other Arabic astronomer, even though there were at least 57 astronomers in the Islamic world after Ibn Shatir, but none of them referred to his work. As stated in chapter 6, Ibn Shatir was employed as the official time keeper (*muwaqqit*) at the Umayyad Mosque in Damascus, his great work on astronomy was a sideshow which was never given much attention by anyone. As *muwaqqit*, Ibn Shatir worked alone and had no student to continue his work. There was no institution or environment to recognise, disseminate, develop further and apply new scientific knowledge.

The works of Ibn Rushd and Ibn Khaldun were not followed up either – in fact they were forgotten, as was the work of Ibn Haitham. Likewise was Ibn Nafis (1210-1288), his discovery of pulmonary blood circulation (chapter 8) was not pursued.

9.3.3 Funding Support for Science

Most Arabic scholars carried out work on philosophy and science as a pastime, but all of them required royal patronage, if not for their livelihood but for their lives – to protect them from the fury of the orthodox. Nearly all scholars had to earn their living by doing other jobs, such as physicians or court officials, despite the outstanding contributions they made. Their list includes: Kindi (d 870), Razi (d 925), Farabi (d 950), Ibn Sina (d 1037), Ibn Haitham (d 1039), Biruni (d 1048), Baghdadi (d 1152), Ibn Rushd (d 1198) and Ibn Shatir (d 1375). The Maragha Observatory (section 6.2) was built by the state, as were many other major observatories, but the astronomers depended on direct patronage. Once the patron died or withdrew support, they became victims of the orthodox.

All of these achievements in science can be seen as being the result of personalised patronage, without any institutional funding. This was not a stable basis for the growth of science in a community. As stated before, in the early period the patron-houses provided a forum for the spread of knowledge, but this facility disappeared after the 9th century due to the pressure from the orthodox.

9.3.4 Inadequate Islamic Law

Sharia never developed the concept of jurisdiction of the court, of writ against the law-enforcers to account for their actions or of any rules of evidence, let alone judicial reviews of government policies. Since all Muslims are members of the same umma, and since every court has jurisdiction over the whole umma, any person could be judged by any qadi of the right madhab for any allegation. The qadi or a ruler handling law cannot be summoned before a court to justify his legal action. The qadi had the sole authority to accept or reject anyone as a witness on grounds of piety. He could ruin the social position of any scientist by declaring him unfit as a witness, which would be tantamount to denunciation as a lesser heretic; but to the masses, heretic just the same, as they could not distinguish between the two. There was no mechanism to challenge unfair application of the law, nor did the law protect the rights of a scientist (as an individual) or of scientific inventions.

9.4 Summary and Conclusion

While the pursuit of religious science gave way to secular science in Europe, it did not do so under Islam; because, with the fall of the Mutazilites the state (which could potentially be secular) lost power permanently to the religious authority (the ulama). The distaste of the orthodoxy for rational knowledge reached a particular high point in the 13th century when they denounced even the Asharite theologians of *kalam* (which included Imam Gazzali) for their use of logic and rational arguments. A.I. Sabra [Sab76], whom we have quoted before, argues that the progress of science continued under Islam right up to the 13th century, but that science was "Islamised", that is, a special view of science was taken "which confined scientific research to very narrow and essentially unprogressive areas". That special view permitted the cultivation of only utilitarian science within a strict religious context.

By the time the thirteenth century was reached, the intellectual climate that had been created by the early Abbasids in the tenth century, gave way to a narrow, rigid and closed society, in which the search for secular knowledge was stifled. The Ottoman rulers heralded their legitimacy on being the champions of the orthodoxy. Even Ibn Khaldun, "the most learned man of his time", the father of sociology, castigated philosophy and declared science to be abhorrent. He stated the philosophers to be "scholars whom God has led astray", and ".... problems of physics are of no importance to us in our religious affairs or in our livelihood" [Sau63]. But such an opinion did not save him from the ever-demanding scrutiny of the orthodox, who found reasons to denounce him for heresy anyway, this time for his social theory of *asabiya* as mentioned earlier.

The orthodox denounced the study of science, logic, mathematics and philosophy as ungodly, because (according to them) these subjects not only contained useless knowledge, but also made one disrespectful of religious laws. Any study for the systematisation of knowledge, including even of grammar, was also distrusted. If a scholar was practising such undesirable topics, their social reputation, and sometimes even life and property could be at risk. Under these conditions, would it have been possible for Ibn Shatir to propose a heliocentric planetary system – an idea originally proposed by Aristarchus of Samos (c 310-230 BCE) as mentioned in chapter 1? Ibn Shatir, a mere *muwaqqit*, could not possibly have allowed such an ungodly thought to develop in his mind, even if the possibility had occurred to him. Perhaps he himself would have dismissed it as a temptation from Satan to mislead and misguide him. Thus, Arabic science failed to make the transition "from the closed world" to "the infinite universe" [Huf95/p68].

Let us look at it from a different angle. In trying to explain why the Inca craftsmen did not invent anything, the eminent scholar Jacob Bronowski said [Bro73/p103] "But naturally, if you work for an Inca (if you work for any one man) his tastes rules you and you make no invention". We can see here a parallel with the closed world of Islamic science that prevailed at that time. The task of a *muwaqqit* (or for an Islamic astronomer) was to determine prayer times, and therefore as long as the model gave an accurate prediction of the times (as it did), the utilitarian need was satisfied; there was no need to improve the model, certainly not for the study of "useless science". Herein perhaps lies the explanation why Islamic astronomers after Ibn Shatir did not work on his model for further improvement. In this sense Ibn Shatir's work defines the boundary of utilitarian Islamic science of the day.

9: CAUSES FOR DECLINE

On the other hand, where could astronomy as a science have proceeded after Ibn Shatir, except to a heliocentric planetary model – an unthinkable idea that would have been condemned by the ulama instantly. A Muslim mind in that closed orthodox atmosphere could never have ventured such an impious thought. Recall what Imam Ahmad Ibn Hanbal said: *Even to discuss a thing that the Prophet did not discuss is an error.* This view of his was the prevailing social norm. Rationalism and causality were all an anathema to the ulama and to the populace. Society treated scholars as ungodly and undesirable. The orthodox focussed on God, to the exclusion of His creation – man and the world in which man lives. Under these conditions Islamic science simply withered away.

> "Arabic science was dead by 1500, there has been no revival under the Ottomans. ... From the sixteenth century onwards, Islam did not simply rest on its oars, it drifted backwards ..". [Sau63].

In the next century, when Europe had produced intellectual giants – like Descartes, Galileo, Newton and Voltaire – Islam had nothing to offer to the world. For the first time in 800 years, the West got the upper hand with a feeling of superiority, it was the time when the West started to pour unanimous contempt openly in their writings for the "sloth, tyranny and backwardness" of the Muslims. In 1670, Jean Chardin, an English traveller of Persia and India was surprised by the lack of intellectual curiosity, technical capability and economic activity in those countries. Worse was to follow. The Muslim great powers of the 17th century, the Ottomans, Mughals and Safavids, started a speedy decline. Turkey went downhill fast after their reverses in Vienna (1683), the Mughal Empire started disintegrating after the death of emperor Aurangzeb (1707), and the Persian Empire of the Safavids gave way (1722) to the destructive Nadir Shah. Islamic powers fell all over the world, never to rise again. But why? That is what we shall explore further in chapter 11.

References and Sources

[Ahm98] Akbar S. Ahmad: Postmodernism and Islam: Predicament and Promise, Routledge, reprinted 1998.

[Ber99] Francois Bernier: Travels in the Mughal Empire (AD 1656-1668), translation from French by V. A. Smith, 2nd Ed, Low Price Publications, Delhi, reprinted 1999, pp155-61.

[Bro73] J. Bronowski: The Ascent of Man, Book Club Associates, 1973, based on the celebrated BBC TV programme of the same title.

[Gol81] I. Goldziher: "Attitude of Orthodox Islam Towards Ancient Sciences", Vol I, Studies in Islam, ed Marlin Swartz, Oxford University Press, 1981, pp185-215.

[Hab79] Irfan Habib: Changes in Tech in Medieval India, paper presented at the Symposium on Tech and Soc, Indian History Congress, Waltair, 1979.

[Hoo91] Pervez Hoodbhoy: Science and Islam, Zed Books, London, 1991.

[Hou83] Albert Hourani: Arabic Thought in the Liberal Age (1798-1939), Cambridge University Press, 1983.

[Hou85] G. F. Hourani: Reason and Tradition in Islamic Ethics, Cambridge University Press, 1985.

[Huf95] T. E. Huff: The Rise of Early Modern Science, Chapter 2, Cambridge University Press, pbk edition 1995.

[Lew73] Bernard Lewis: Islam in History, Alcove Press, London, 1973.

[Lew94] B. Lewis: The Muslim Discovery of Europe, Phoenix, 1994.

[Lew97] B. Lewis: The Middle East, Phoenix Giant, 1997, [Arms from Europe: pp275-77].

[Qua00] Donald Quataert: The Ottoman Empire, 1700-1922, Cambridge University Press, 2000.

[Rah79] Fazlur Rahman: Islam, University of Chicago Press, 2nd ed. 1979.

[Rah84] Fazlur Rahman: Islamic Methodology in History, Islamic Research Institute, Pakistan, [1965, 1984].

[Rut91]: Malise Ruthven: Islam in the World, Penguin 1991.

[Sab76] A. I. Sabra: Ch 7: "The Scientific Enterprise", Islam in the Arab World, ed. B. Lewis, published by Alfred A. Knopf, New York, 1976, pp181-200.

[Sau65] J. J. Saunders: A Medieval History of Islam, Routledge, 1965.

[Sau63] J. J. Saunders: "The Problem of Islamic Decadence", Journal of World History, Vol 7, 1963, pp701-720.

[Say86] Sayili, Aydin: Turkish Contributions to Scientific work in Islam, 1986.

[Sch79] Schacht and Bosworth (Editors), Legacy of Islam, 2nd ed, Oxford University Press, 2nd ed. 1979.

[Sea03] R. Segal: Islam's Black Slaves, Atlantic Books, London, pbk 2003.

CHAPTER 10

GREATEST "HERETICS" AND THEIR NEMESIS

We present here biographies of six most outstanding intellectuals of Islam, all of whom were declared heretics by the Muslim orthodox, some posthumously. In narrating their lives, we shall cover the social environments in which they lived and worked, the contributions they made and the brilliant legacy they left behind, not forgetting their tenacity, persistence and struggle in achieving what they achieved. Their nemesis was Imam Gazzali, the greatest anti-intellectual intellectual of Islam, one who denounced every Greco-Islamic philosopher and secular scientist as heretic and who is often credited with building the intellectual foundation for banishing rationalism from Islam. The forces of anti-rationalism he espoused gained in strength rapidly and overtook even him eventually, labelling him also as a suspect, for his appreciation of logic.

The six so called "heretics" we shall consider in this chapter are:

- al-Kindi (801-873)
- al-Razi (865-925)
- al-Farabi (870-950)
- Ibn Sina (980-1037)
- Ibn Rushd (1126-1198)
- Ibn Khaldun (1332-1406)

We have also included a note on Yahya Suhrawardy, a martyr who extended Ibn Sina's ideas to sufism. The life and work of Imam Gazzali (1058-1111), their scourge, is presented last.

Observe that there was a gap of nearly 100 years between Ibn Sina and Ibn Rushd and more than 100 years between Ibn Rushd and Ibn Khaldun, signifying a slow decline in the number of great scholars as time passed. These scholars, except for Ibn Khaldun and Gazzali, worked in Greek-based Islamic philosophy, often referred to as *falsafa* (or *falsafah*). Therefore we shall provide below a brief introduction to Greek philosophy and its relationship to falsafa, before we attend to the biographies.

Three Greek giants who exerted major influences on Islamic (as well as Jewish and Christian) philosophy were Plato (429-347 BCE), Aristotle (384-322 BCE) and Plotinus (205-270 CE), each with a different approach. Plato's view was that the world was created by God at a moment of time out of some kind of formless chaotic matter. Therefore the world was not eternal, nor was it created out of nothing. The act of creation produced order on the pre-existing chaotic matter. Razi accepted this view of order, but other Islamic philosophers did not.

According to Aristotle, who was opposed to his rival Plato, the world (i.e. the universe) is eternal in which the heavenly bodies are in eternal and constant motion and the earth (which is a part of the world) is a constant rotation of forms over matter. This eternal process could be changed only by the eternal God as *the prime mover* (i.e. *the first cause*). God is an intellectual principle, *thought thinking thought*, remote from the individual concerns of men. Therefore in Aristotle's view the world (like God) was eternal, and hence the world was not created by God at a moment in time.

Some 600 years after Aristotle, Plotinus, who lived first in Alexandria and then in Rome, formulated the new philosophy of Neoplatonism, re-interpreting the ideas of Plato on God and creation. In it God is referred to as the *One*, the source of all existence. The *One* is totally other than all else, transcending Aristotle's concept of God. Furthermore the *One* is not only the first cause (which keeps the world in perpetual motion), but the *One* is also the source from which all existents emanate, like the light emanating from the sun. Emanation is an eternal creative process. Plotinus also subscribed to the concept of *soul* that controls the body.

There were differences between the Islamic and Greek concept of God and His creation [Arm01/p60]. In Islam, God created the universe, He concerns Himself with earthly events, and He will judge men at the end of time. Unlike the Muslims, the ancient

Greeks did not believe that God concerned Himself with earthly events, or that He would judge men at the end of time. However, Plato thought that God created the universe at a moment in time out of chaos, while Aristotle viewed the universe as uncreated and eternal. In contrast, Plotinus, who was closer to Plato than to Aristotle, conjectured the universe as an *emanation* from *One* God, at a moment in time.

Despite some of these differences, the Muslims found the central concept of God in the Greek philosophy very useful in their efforts to understand God through reason. They identified the *One* of Plotinus with the God of Aristotle, and argued that this *One* was the same as the God of the monotheistic religions. They also found the doctrine of emanation very attractive as it helped them to explain how God was totally dissimilar to His creation but yet could create something that did not resemble Him in any manner. Therefore they accepted the idea that all things emanate eternally from changeless eternal God, Who Himself lies beyond that complex entangled emanation.

Furthermore, the Muslim philosophers came to view that God was the Absolute Intellect (or Reason) and human intellect, a divine gift of reason, was a reflection of that Absolute Intellect. With this divine gift, a man could purify his mind of all that which was not rational, and could learn to live in a reasoned way. By following this path of reason a man could reverse emanation away from God, rise above the complexity of earthly life and get closer to that Absolute Intellect, the ideal simple Truth of *One* – the Truth of the *tauhid*.

They also believed the revelations of the Quran to have both outer (*zahir*) and inner (*batin*) meaning. The outer (or external) truth is intended for common people, but the *batin* truth, which is the greater truth, is beyond the understanding of ordinary people. It can be discerned only by rational reasoning, which the intellectuals can carry out. This greater truth can also be perceived, in the absence of a revelation, by rational reasoning alone[1]. In this sense, they placed reason above revelation, which attracted the wrath of the orthodox. And yet, they were devout Muslims, who held rationalism to be the highest form of the religion of Islam.

---

[1] This is an important theological principle that enables the use of rational reasoning to conduct ourselves, such as to make laws, in areas not covered in the Quran.

All Muslim philosophers, partly following their Greek predecessors, assumed the earth to be the centre of the universe, all the planets (including the sun and the moon) circling round it in different layers, with stars forming the penultimate layer, before a final empty layer. And man inhabiting this earth is the supreme creation of God, as he is endowed with reason (intellect). Apart from belief in this structure, the Muslim philosophers differed on their approaches. Kindi believed in the alternative doctrine of *ex nihilo* creation, which Razi did not, but both believed in the basic emanation principle. Farabi unified the idea of Aristotle's God with the *One* of Plotinus, as the God of Islam, and developed an intelligence-based emanation scheme to explain all things, even human society. Ibn Sina refined Farabi's emanation scheme, with the introduction of a concept of essence. Ibn Rushd, the most rationalistic Muslim philosopher, also refined Farabi's approach, but with a more Aristotelian flavour, particularly rejecting some special ideas of Ibn Sina.

10.1 Abu Yusuf Yaqub al-Kindi (801-873)

Abu Yusuf Yaqub Ibn Ishaq al-Kindi, known as the philosopher of the Arabs, was born probably at the start of the 9th century CE in Kufa where his father Ishaq Ibn Sabbah was the Governor. He was from an Arab aristocratic family whose members held important public positions; an ancestor was a Companion of the Prophet and further back the ancestors were South Arabian Kinda kings. Abu Yusuf was educated first in Kufa and then in Baghdad, acquiring a profound knowledge of Greek, Persian and Indian arts of wisdom. A man of means, he was associated with the nobility and the rich, and probably provided financial assistance for translations. He was the kind of philosopher, a Mutazilite, Khalifa Mamun admired in his Baitul Hikmah. He climbed great heights and received the patronage from three Khalifas: Mamun, Mutasim and Wathiq, later becoming the tutor of Mutasim's son Ahmad.

It would seem he was most interested in natural sciences, but he had also carried out work in logic and metaphysics. He believed that philosophy could not be grasped except through mathematics. It is sometimes claimed that he was the first person to introduce Aristotelian thought into Islamic study. He considered metaphysics to hold "the highest honour and rank ..., because the science dealing with the cause is more honourable than the science dealing with the caused". He believed truth to be supreme and universal, which scholars must seek to complete and contextualise, within different languages and customs. This is just what is proposed by today's *ahl al-Quran* reformers (section 13.3). Like other philosophers, he believed in two forms of truth, one for the uneducated masses and the other for the educated. The first group can understand only simple things, such as *houris* in the Heavens, while the second group was endowed with logic and reason, and therefore could perceive a deeper truth. Any passage in the Quran that conflicted with reality, he argued, must be treated as allegorical so that the deeper truth contained in it can be discovered, as shown in the following example. Consider the verse, Quran [55:5], which says something like:

... the sun, moon, stars, trees 'prostrate themselves'
before God ...

Reading this verse, the uneducated mass will have a vision of a physical prostration before God, while the educated can interpret 'prostrate' as 'submit', implying universal submission to God's will. With further abstraction, an educated person can deduce even the deeper truth of universal laws of God. Thus according to Kindi, allegories allow men of education and reason to discover more profound truths, which cannot be perceived by the masses. This is why he defined his philosophy (*falsafa*) as "knowledge of things as they are in reality, according to human capacity". Accordingly, he believed revelations to be meant for all men, but their meanings to be different to different men, depending on their intellectual level. He declared that truth was universal and supreme, and that the truths of religion and philosophy were in accord.

Contrary to Aristotle, he viewed *time* and *movement* to be transient (that is, not eternal) and finite. If time and movement are finite, he argued, then the creation (which is a form of movement in a finite time) is finite; all finite things are created by a Creator (God) who alone is eternal and infinite. He believed God to be the prime cause, since otherwise the causation would continue indefinitely, which he concluded would be absurd.

Kindi was a polymath, and wrote some 290 books on topics ranging from philosophy, astronomy, mathematics, to cryptography (see section 5.5), and from medicine to linguistics and music. However, only a few of his books have survived. He innovated the idea of, and played a pivotal role in, applying Greek thought to the understanding of Islam. His discourses would have been attended by members of the Khalifa's family, other aristocrats, fellow scholars, students and theologians. His sophisticated arguments were not always well-received particularly by the theologians, let alone by ordinary men, whose intellectual capability he despised, as did other philosophers of his time. He was also known as the philosopher of the Arabs [Fak83/p69], as he was the only significant philosopher of Arab descent, unless one counts Ibn Khaldun (over 500 years later) as a mainstream philosopher of theology, which he was not.

However, the age of reason died with Khalifa Wathiq. His successor Khalifa Mutawakkil ended the long period of liberalism, that was initiated by Khalifa Mansur, and unleashed rampant Sunni orthodoxy, which declared all such philosophers as being a danger to Islam. The brightest jewel in the most advanced cultural centre of the contemporary world, one that served three Khalifas with great distinction, became the first major victim of the Sunni wrath. Khalifa Mutawakkil declared Kindi a heretic when the latter was sixty. His personal property was confiscated, and his great private library, well-known as *al-Kindia*, was set on fire (though some books were saved by a sympathetic friend through subtle bribery). In addition, this old man was administered 50 lashes in public as further punishment, the public jeering at every lash with approval, according to an eye-witness report. This public flogging broke him, even though he viewed suffering and death as an inescapable part of human destiny. He retired to his home, where he, a devastated old man, suffered in silence and with dignity until his death at the age of 72 in 873 CE. He paid with his life for the love of truth as perceived by his brilliant mind. Such was the reward he received from the Muslim conservative society for his intellectual achievements.

10.2 Abu Bakr Mohammad al-Razi (865-925)

Abu Bakr Mohammad Ibn Zakariya al-Razi was born in Rayy, near modern Tehran, in 865 CE and died in 925 CE. He is known as the clinical genius of the Islamic world, the 'Arab Galen', but he was also a philosopher who was fearless in expressing his thought. A poet, singer and musician in the early youth, he left Persia to study medicine in Baghdad, where he later became the head of its famous Muqtadari hospital. Although he lived a luxurious life, he is said to have treated both rich and poor patients equally and with extraordinary care. He moved around various cities from time to time, finally settling down at Rayy where he earned fame and also attracted the wrath of the orthodox, both in equal measure, before he died blind. In *falsafa*, he went furthest on his journey to reason and rationalism, as he placed reason above revelation, as is explained below.

He had written around 200 treatises, of which about half were in medicine and 21 on chemistry. Some 40 of his manuscripts are still extant in museums and libraries. His important medical contributions include (i) *Kitab al Hawi* (continens) – an enormous encyclopaedia on the Greek, Indian and contemporary Arabic medicine with his personal comments from his own experience, (ii) *Kitab al-Mansuri* – a smaller collection of ten books on Greek science, and (iii) *Kitab al Jadari wal Hasba* – his monograph on small pox, measles and other diseases (see also chapter 8). These have been translated into Latin and other European languages. Many contributions to gynaecology, obstetrics and ophthalmic surgery can also be traced back to him. He was a polymath, with contributions to philosophy, astronomy, mathematics and physics. In chemistry, he was the first chemist to produce sulphuric acid, and in physics he studied specific gravity, by means of hydrostatic balance.

Only a fragment of his vast philosophical work has survived, which does not give a comprehensive picture of his philosophical doctrine. Unlike Kindi and Farabi, he did not believe that religion and reason could be reconciled. He championed Plato over Aristotle with a deep admiration for Socrates and developed a philosophical system (drawn from some Platonic ideas) in which God is seen as being related in a complex manner to soul, matter, space and time. In his religious view, God has endowed humans with intellect, thus enabling them to understand the universe, conduct their own lives in a way that achieves salvation, without in principle requiring any prophet (or revelation) for guidance. In this sense he had placed reason above revelation, which infuriated the orthodox. He also said "The prophet is above the philosopher .... for the prophet is delegated and the philosopher is delegated unto him". Some later Muslim writers condemned him for speaking openly on the superiority of reason above revelation. Even the scientist al-Biruni, possibly to please the orthodox, denounced Razi and attributed his later blindness (see below) to divine punishment – a despicable comment from such an eminent fellow scientist, who was expected to hold rational opinion. Most of his philosophical works were destroyed by the orthodox, following their well-established technique of setting fire to whatever they did not approve of.

It is said that the amir of Bukhara, enraged at his heresy, ordered his minions to hit Razi on the head with his own book, until either the head or the book cracked. His head did not crack, but he became blind in both eyes. Later when a remedial eye surgery was suggested, he replied: "I have seen enough of this world, and I do not cherish the idea of an operation in the hope of seeing more of it", as quoted in [Hoo91]. He died soon afterwards. This greatest man of Arabic medicine was thus rewarded for his freethinking by the society he served all his life.

10.3 Abu Nasr al-Farabi (870-950)

While Kindi broke the ice on Greek thought and pioneered the idea of harmonising Quranic teaching with systematic rational metaphysics, it is Abu Nasr al-Farabi who produced the first real philosophical model (*falsafa*). He was born in about 870 CE in a village called Wasij in Transoxania and died in Aleppo (Syria) in 950 CE. His father was a general of Turkic origin, and he himself served as a judge for some time. An eminent logician, he devoted himself to intellectual studies, aloof from the political turmoil and intrigues of the time. Information on his life is scanty. He did not leave any autobiography (as did Ibn Sina and Ibn Khaldun), nor did any of his immediate pupils write about him, but his voluminous writings influenced many generations of scholars, even long after his death.

Apart from the Quran and hadiths, he studied linguistics and jurisprudence, and learnt Arabic, Persian and Turkic. He then moved to Baghdad where he studied logic and became the foremost logician of his time, earning the title "the second Aristotle". After spending 20 years in Baghdad, he moved to Aleppo to join the scholarly court of Shia'ite sultan Saif al-Daulah. Despite the pro-Arab sympathy of that court, Farabi, the Turk, felt completely at ease there, devoting wholly to intellectual pursuits and shunning the glamour of the court life. Living and dressing like a sufi, he continued to produce his written works, sitting under the shade of thick green foliage, near murmuring streams. He believed religious and philosophical truths to be indivisible and created a theory to integrate them through logic and reason.

He believed in rationality and causality (denied by Gazzali and the orthodox). Like Plato's Republic he produced a concept of philosopher-king and also a concept of a model (organic) city. The city was to be organised into parts, like in a body. If one part goes wrong, other parts can react and take care. Social activities and individual duties (according to abilities) are organised and distributed to meet the social needs, harmony and well-being. The Chief must be stout, super-intelligent, lover of knowledge and supporter of justice – sounds like an ideal Shia Imam.

Farabi's writings were always precise, brief and to the point – a typical characteristic of good logicians. But these were esoteric ideas with high intellectual content, not meant for the masses, for whom he, like all his contemporary philosophers, had contempt. His work became available in the West during the 10th and 11th centuries, when some Andalusian scholars became his disciples. The translation of his writing into Latin appeared earlier than those of other Arabic scholars, implying a major impact on Western thought. He has influenced both the Jewish and Christian scholars, including even the celebrated Jewish philosopher Maimonides, who owed much to Farabi. Some of his writings were translated into Hebrew and Latin right up to the 19th century, while translations into modern European languages are continuing even today. The strict logician Ibn Rushd was influenced by Farabi's theory of intelligence, while Ibn Khaldun embraced and expounded it. However, Farabi favoured science, advocated experimentation, affirmed causality, and denied astrology – a very modern approach. He elevated intellect to a higher level with a view to reconciling philosophy and religion.

Perhaps because of his aloofness from politics and power, unlike Kindi, and perhaps because of his solitary living in Aleppo, far away from the zealots of Baghdad, he avoided persecution by the orthodox during his life, but they nevertheless caught up with him after his death. He was posthumously branded by the later orthodox (including today's) for applying pure reason, rather than Sharia, to distinguish good from evil. To the orthodox, freethinking is a crime even today.

10.4 Abu Ali al-Hussain Ibn Sina (980-1037)

The tenth century in Persia (particularly after 945 CE) was a turbulent period, when Abbasid power waned, giving rise to a number of feuding independent sultanates (who nevertheless recognised the weak Khalifa in Baghdad as the head of the Muslim umma – chapter 2). This was also the time when the Persian renaissance started to bloom, with prose and poetry being written in Persian rather than in Arabic. Many Persians devoted themselves to the pursuit of science and philosophy, and one such Persian was a genius named Abu Ali al-Hussain Abdullah Ibn Sina, who was born near Bukhara in 980 CE into a Shia family.

His father was a governor of a village under the Samanid sultan Nuh Ibn Mansur of Transoxania and Khurasan (which included Bukhara and Samarqand). Ibn Sina was mostly self-taught, as he stated in his autobiography, supervised by his father whose home was a meeting place of the local scholars. From his childhood Ibn Sina displayed a prodigious memory and remarkable intellect, as noted by those scholars who visited his father's house. By the age of 10, he had memorised the whole of the Quran and most of the Arabic poetry that he had read. At 13, he started studying medicine and mastered the subject by the time he reached 16. He then began treating patients. He received instruction only on logic and metaphysics, while learning other subjects by self-study. After he cured the sultan of an illness, he was given free access to the royal library, which proved exceedingly valuable in quenching his immense thirst for knowledge.

Following the defeat of the sultan in a war, Ibn Sina lost his patron, and wandered around in different parts of Khurasan for five years, working as a physician and administrator by day and as a teacher of philosophy and science by night. In Hamadan (western Iran) he became a friend of the Buyid amir Shams al-Dowla, who appointed him twice as his vizier. He however continued his scholarly studies with an independent mind, insisting on the primacy of reason, although he rejected the Mutazilite doctrine. In one of his writings he explained how he had handled intellectual challenges – going to the mosque and praying until God revealed a solution, then working hard throughout the night and finally taking a sip of wine to relax. His rationalism, coupled with that sip of wine, was all too much for the orthodoxy. They denounced him as a heretic, and after a heated philosophical argument with him, the orthodox army decided to solve the Sina problem for good by eliminating him physically. But not finding him in his house (he was tipped off), the angry army plundered his house and demanded that the amir behead him.

Ibn Sina went into hiding, but was caught later and imprisoned in Hamadan.

All through this turmoil, he refused to compromise his belief. When advised by friends to moderate his stance on the supremacy of logic, he replied: "I prefer a short life with width, rather a narrow one with length". In 1022 CE he escaped from prison to Isfahan where, he entered the court of sultan Ala al-Dawla and lived there peacefully until his death in 1037 CE. In Isfahan, he carried out some astronomical observations, leading to some new findings, such as, that the planet venus is closer to the sun than the earth is. Many of his works were composed during military campaigns, when he had to accompany his patron the sultan. During one such campaign in 1037, he fell ill and self-administered some medicine, but did not recover.

His two most important works, *the Book of Healing* and the *Canon of Medicine*, were completed in Isfahan (see chapter 8 on Medicine for more details). The first is an encyclopaedia covering many branches of science, such as astronomy, arithmetic, geometry, logic, natural science, music and psychology. The second is the most famous single book on medicine, comprising some ten million words (about 20,000 A4 pages[2]), on which he started to work while in hiding in Hamadan. He produced about 450 books and articles, of which 240 have survived, including 140 on philosophy and 40 on medicine.

Another area of his major contributions was philosophy in which he discussed knowledge and reality, and tried to construct a rational foundation for understanding God and religion, based on a Neoplatonic approach, with a view to integrating all aspects of science and religion in a grand metaphysical vision. God was the pure intellect and knowledge consisted of the mind *grasping* the intelligible, with reason and logic. In his vision, a prophet was superior to a philosopher, since he did not have to depend on human reasoning, as he enjoyed a direct and intuitive knowledge of God. He also wrote in

---

[2] Ponder for a moment the meaning of 20,000 A4 pages. I estimate that the enormous volumes written by George Sarton on the history of science [referenced in this book in many chapters] come to about 8000 A4 pages altogether. Most of us, science-academics all our lives, could never produce even 5000 of such pages on weighty research in technical topics. Contrast this with Ibn Sina who never had a cushy academic job. This is just one book, he wrote several hundred other books, though of smaller sizes.

favour of natural sciences and against the ulama who, in his view, attempted to obscure the truth by objecting to the study of natural sciences.

Falsafa, initiated by Kindi and improved by Farabi, reached its apogee in the work of Ibn Sina, who produced a rational demonstration of God's existence based on Aristotle's proof, that became standard among later Jewish and Christian philosophers, including Roger Bacon[3] (d 1204) and St Thomas Aquinas (d 1274).

It would seem that in his later life Ibn Sina leaned more towards mysticism, and wanted to relate his rationalism to the religious experience of sufism, and might have written a book on it. This relationship with sufism was finally formulated by Yahya Suhrawardy, a follower of Ibn Sina and intellectually as rigorous as Farabi. Suhrawardy who placed intuition above reason, in combining the philosophy of Ibn Sina with mysticism (sufism), was put to death in Aleppo in 1191 for his ideas. Aleppo was no longer far away from the orthodox, as it was in Farabi's time.

In later centuries the orthodox treated Ibn Sina as a heretic. Imam Gazzali declared him an unbeliever for 'transmitting the philosophy of Aristotle', which crystallises the outlook of the orthodox even today. Such was the price paid by intellectuals under Islam. We conclude this section with a poem by Ibn Sina, composed to defend himself against the charge of heresy:

It is not so easy and trifling to call me a heretic
No belief in religion is firmer than mine own
I am the unique person in the whole world if I am a heretic
Then there is not a single Mussulman any where in the world

as quoted by S. H. Nasr in [Nas78/p183].

10.5 Muhammad Ibn Rushd (1126-1198)

Muhammad Ibn Rushd, one of the greatest Muslim philosophers, was born in Cordoba in 1126 CE in a family of distinguished jurists and theologians. At the age of 42, he joined the court of the Almohad ruler Abu Yaqub (1163-1184) at Cordoba, who commissioned

---

[3] Unfortunate Roger Bacon who did not know that he too, like his Muslim hero, would later suffer in prison for his impiety to his religion, in his case Christianity.

him to write a summary of Aristotle's work for his clearer understanding.

The next 26 years he spent producing detailed commentaries on Aristotle, defending his philosophy against attack by the orthodox, and constructing a form of Aristotelianism, largely cleansed of Neoplatonism. This work later earned him the title of "The Commentator" in the Latin West. As a philosopher, he tried to integrate Aristotelian philosophy and Islamic thoughts, asserting that truths cannot contradict each other, and therefore there is no incompatibility between religious and philosophical truths, when understood properly. However, he regarded philosophy, rather than religion, as a better framework for acquiring knowledge. Unlike the other philosophers, he was both humble and generous, one who never desired power or possessions.

He did not write much on theology, but he criticised both Farabi and Ibn Sina on some theological points, lashing out in particular at Gazzali against the latter's denial of causality. According to Gazzali (section 10.7), we cannot say fire burns a ball of wool, because it will burn only if God wills it to burn at that occasion. Ibn Rushd ridiculed this denial of cause and effect. In fact he wanted philosophical arguments in Islam to be pursued and developed further to attract intellectuals who would strengthen the faith, while recognising that for the vast majority (the ordinary people) a simple belief was what mattered. This view of his was interpreted by the orthodox as placing philosophy above religion and hence it was declared as heretical.

Ibn Rushd attempted to resuscitate *falsafa* from Gazzali's devastating polemic, championing the cause of rationalism against the blind faith of the orthodox, but it came too late. His work was not followed up by the later Muslims. Nevertheless, he is the best known Arabic philosopher in the West and he stands out as one of the leading international scholars of all time. However, much of his work that is available exists only in Latin and Hebrew translations, the original Arabic text having been lost long ago. It is his work that gave Western scholars the first substantive introduction to Greek thought, and his philosophy along with that of Ibn Sina provided the basic inspiration to later Western scholars such as Duns Schotus, Albertus Magnus, St Thomas Aquinas and Roger Bacon. But it did not endear him to orthodox Christians, who joined Muslims in burning his books – a rare instance of Christian/Muslim unity!

However, work on philosophy was not his real job. He was appointed the qadi of Seville in 1169 CE and two years later he succeeded his father as the chief qadi of Cordoba. In 1182 he also became the court physician. He continued to serve the Almohads until he fell into disgrace four years before his death. When Abu Yaqub died in 1184, Ibn Rushd lost his patron; the successor Abu Yusuf (1184-99) first banned Ibn Rushd from studying logic and science, and then banished him from Cordoba to North Africa along with other students of philosophy. Ibn Rushd was then 70. His books were burnt. He died in 1198 in Marrakesh (the capital of the Almohads) after being officially "rehabilitated", but still being suspected as a heretic by the orthodox.

10.6 Abd al-Rahman Ibn Khaldun (1332-1406)

The 600th anniversary of the birth of Ibn Khaldun, the great Muslim genius, was celebrated in 1932. But Muslims had ignored and lost him for centuries until he was rediscovered by the West in the 19th century as one of the greatest philosophers of Islam. No less a person than Arnold Toynbee (1889-1975) the great British historian commented, that Ibn Khaldun "conceived and formulated a philosophy of history which is undoubtedly the greatest work of its kind that has been created in any time or place". He is also recognised today as the father of economics, anthropology, political science and historical geography, but his ideas were not absorbed by his society, nor were they carried forward by its future generations. At that time the Islamic empire had passed its peak, and Ibn Khaldun disappeared from the Muslim view, much to the satisfaction of the orthodox.

Born in Tunis in May 27, 1332, Abd al-Rahman bin Muhammad Ibn Khaldun of Arab ancestry from Yemen, was educated by his father who was a scholarly person. His ancestry had a long involvement in North African politics at the highest level, with some of his relatives holding high positions, hiding, fleeing, and occasionally reigning. He memorised the Quran, studied the traditional religious material, and learnt jurisprudence, philology and poetry. When he was 19, the great plague struck North Africa, killing his father.

Young Ibn Khaldun entered into the services of various sultans in Morocco. His own fortunes waned and waxed with those of his patrons, but in the process he also made friends and enemies in high places. He entered into political conspiracies with some of them, often changing loyalties, for which he was even once imprisoned. However, he always found time for advanced studies, which he completed, taking advantage of the scholarly environment in Fez. He later left Morocco for Andalusia, where after a successful conspiracy, the victorious sultan Abu Salem rewarded him with the post of secretary. Two years later he was appointed the chief justice, a post that he lost and gained several times following the overthrows of sultans and also due to political intrigues. He returned to Tunisia, his birth place, after 27 years and then moved to Cairo where he was warmly received by the scholars. He became a Maliki (the predominant madhab in Egypt) judge in Cairo. Fourteen years later, after many ups and downs, he became the chief Maliki judge. He accompanied sultan Faraj of Egypt to Damascus where he successfully negotiated for his sultan a treaty with Timur Lang (1336-1405), who admired him. He then returned to Cairo.

During his Cairo period, he studied further and gave lectures at al-Azhar. He completed his famous work *Universal History*, which has survived. He also wrote on theology, and sufism, but some of his books, including one on Timur are lost. He died while still the chief justice, in March 1406, and was buried in the sufi cemetery in Cairo.

His magnum opus *al-Muqaddima* has three parts: *Introduction*, *Universal History* and *History of the Magrib*. The *Introduction* provides the most penetrating contribution of all time on the study of society and state. It covers his historical theories encompassing a vista of perspectives as can be discerned from the topics covered in it: Human society (ethnology and anthropology), Rural civilisation, Forms of governments and institutions, Society of urban civilisation, Economic facts, and Science and humanity. His ideas were formulated directly from his knowledge and practical experience. A statesman and historian, Ibn Khaldun is hailed as the first to attempt a scientific study of the state and society.

He is credited with the understanding of the modern processes of supply and demand on prices and production. In addition he is also credited as the founder of the discipline of anthropology. His historical theories include political and social cycles, and the behaviour of social groups (*asabiya*), including the behaviour of the conqueror and the conquered societies. He rejected occult sciences, such as magic, alchemy and astrology, as pseudo-science. But he also objected to the study of physics as being unnecessary (as mentioned in chapter 9) following the prevailing utilitarian norm of the day.

Ibn Khaldun was nevertheless a major protagonist of anti-Greek thoughts in Islam, as was Imam Gazzali. He rejected the Neoplatonic approach of Farabi and Ibn Sina as unacceptable on religious grounds. He did not believe in the application of reason for the understanding of faith. But this was not enough to save him from the orthodox who by then were more powerful and even more demanding. Their first objection was his rationalism, but worse to them was his concept of *asabiya* (group cohesion) which, while reinforced by a religion, was also meant to be a pre-requisite for the fulfilment of the mission of even a divine religion (that is, prophethood). To say that something is a pre-requisite to prophethood was blasphemy to the orthodox. Ibn Khaldun also outraged the Arabs (he himself was of Arab descent) by attributing glories of the Golden Age to non-Arab scholars and by berating the Arabs for their failure in the field of scholarship, particularly when both their religion and prophet were Arab[4]. He became unpopular with his fellow Arabs for such criticisms. He had been accused not only of heresy, but also of being a "dishonest rationalist masquerading as a Muslim", one whose books should be burnt.

In conclusion, his greatest contribution to Islamic thought was his positivistic philosophy of history and social evolution, and the development of a science of civilisation. According to him, the laws of this science are reducible to a series of geographical, economic and cultural patterns and to some dialectic of historical developments, as elaborated in his Introduction of the *al-Muqaddima*. He developed systematically a complete theory of social determinism, without any parallel until modern times.

10.7 Abu Hamid al-Gazzali (1058-1111)

Imam Abu Hamid Gazzali was the most outstanding scholar of orthodox Islam, who was given the honorific title of *Hujjat al Islam* (Proof of Islam). His intellectual prowess was such that one could understand his opponents' points of view better from his summaries

---

[4] This is what Ibn Khaldun is said to have said: "It is a remarkable fact that with few exceptions, most Muslim scholars both in the religious and intellectual sciences, have been non-Arabs. When a scholar is of Arab origin, he is non-Arab in language and upbringing and has non-Arab teachers. This is so in spite of the fact that Islam is an Arab religion, and its founder was an Arab." [Ros67/p311].

than from their writings. His younger brother Ahmad Gazzali also became a great scholar, though not of the same stature.

Born at Tus (near modern Mashad) in Khurasan, Abu Hamid Gazzali and his younger brother Ahmad became orphans at an early age. Abu Hamid first studied at Tus and later philosophy at Nishapur under the outstanding Asharite theologian Abdul Malik al-Juwayni who was also known as Imam al-Haramayn (d 1085). Abu Hamid achieved such great fame as a scholar, that at the age of 36, in 1084 CE, he was appointed the director of the prestigious Nizamiyyah madrasa in Baghdad by the Seljuk prime minister Nizam al-Mulk Tusi (see section 2.4), with a brief to defend Sunnism against the challenges from the Ismaili Shias, who were then a powerful political force (and who later assassinated Nizam al Mulk). However, Gazzali began in earnest to search for truth – the truth to prove the certainty of the existence of God. He explored *kalam* (Asharite theology), *sufism* and *falsafa* (Greco-Islamic philosophy).

He recognised that certainty itself is a psychological condition which is not necessarily objectively true. He rejected that the Ismaili concept of a *Hidden Imam* could lead to the proof of God's existence, and also concluded that kalam was flawed as it was based on a number of propositions from scripture that could not be verified objectively. While he accepted that the study of reason and logic could provide deeper knowledge, he could not see how this could prove or disprove God's existence. In his treatise *Incoherence of the Philosophers*, he berated both Farabi and Ibn Sina, and rejected reason and logic as a means of finding God. He was thinking so intensely that he went into clinical depression in 1094, joined sufism and gave up his influential teaching post. After practicing sufism for some ten years, he found it to be a way of life that created certain mental states, but did not provide objective proof that of God's existence. Sufis, he concluded, experienced God intuitively, but they could not find God as objective truth verifiable by a rational process. He then came to the conclusion that God could be found only by experience, since that the reality of God lay outside the realm of sense, perception or logical thought; and accordingly he devised a doctrine of belief to serve God. He was later re-appointed to his prestigious Nizamiyyah post, where he worked until retirement and was succeeded by his brother Ahmad.

Physics (or modern science for that matter) cannot prove or disprove the existence of God, as it deals only with observables (things that can be observed experimentally, directly or indirectly). As the Quran [2:02] implies belief in a God is a belief in the unknown, and hence physics cannot know. Assuming God as *the first cause*, the rest of the physical world can be bounded to causality. But Imam Gazzali denied this causality, thus killing off all rational thinking. An intellectual giant though he was, Gazzali created a theology that was irrational, anti-science and anti-intellectual, one that kept Muslims away from participation in rational discourse and scientific advancement (see also chapter 9).

He launched an attack on the rationalist philosophers and their Greek forebears, declaring their approach to be internally incoherent. He, however, differentiated parts of the Greco-Islamic pursuits into those that conflicted with religious truth and those that did not. He accepted logic *not* to be in conflict (and hence it could be studied), but the rest including physics and metaphysics to be in conflict (and hence these could not be studied). His position on mathematics was more complex. In his view mathematics gives clarity and precision, and therefore produces the unachievable expectation of similar clarity and precision in religious doctrine; and thus it weakens a believer's faith. He concluded that mathematics could be a cause for unbelief, and hence it should not be studied.

Imam Gazzali himself studied both logic and mathematics, and appreciated logic as he found it useful in developing his own arguments. But some later orthodox scholars were not so forgiving. Imam Taymiyya (d 1328) launched a vehement attack against Greek logic and its practitioners, which of course included Imam Gazzali (see the criticism of Ibn Salah and others in chapter 9).

In his universe, there is no cause and effect. God is omnipotent and omniscient, unbounded by any universal law, and He does everything on each occasion without any regard to continuity – through a process of arbitrary annihilation and recreation. Events and actions are caused by divine intervention, through annihilation and recreation on every occasion (and hence *the theory of occasionalism*)[5]. In his theology there is no recognition of natural laws based on inherent properties; and therefore, if a fire is set to a ball of dry wool, we

---

[5] This doctrine of Imam Gazzali has *no* relationship to the proposed theory of multi-verses in modern physics, in which new universes are created at every quantum alternative. In any case, all scientific theories are grounded on rationality and causality (denied by Gazzali) – and in addition they are falsifiable, a religious belief is not.

cannot say that the fire will burn the wool on grounds that it is the inherent nature of a fire to burn a substance. In fact, in his belief, the fire will not burn the wool, if God does not will it explicitly. Thus, at each occasion God himself decides, arbitrarily and in His wisdom, what is to be done. To say otherwise is a *shi'rk*, which is the great sin of associating the power of God with some other agencies, violating the doctrine of strict unity (*tauhid*) in Islam, since only God is the agent of any action. Thus, God annihilates and re-creates the universe at every occasion, as the omnipotent Being, unbounded by any constraint. There are no causality or laws of nature. If anyone doubts this theory, then that person doubts God's ability, hence he is an unbeliever, as Gazzali declared the philosophers to be. An added crime of Ibn Sina in the eyes of Gazzali was his non-belief in the resurrection of the same earthly physical body. Ibn Rushd and Ibn Khaldun came later and hence escaped Gazzali's wrath, though not the wrath of the other later orthodox scholars.

Let us briefly consider the consequence of the Gazzali doctrine on science, using an example from Pakistani orthodox thoughts cited in [Hoo91]. Consider the chemical formula of water molecule from hydrogen and oxygen molecules:

$$2H_2 + O_2 \rightarrow 2H_2O$$

If this formula fails even once under test condition, then this formula is falsified (i.e. invalid) and hence must be revised. Scientific statements must always be subject to this falsifiability test. However, in Imam Gazzali's doctrine, the above formula will be a *shi'rk*; we must rewrite it as:

$$2H_2 + O_2 \text{ [if God wills]} \rightarrow 2H_2O$$

What does it mean? Does it mean God sometimes does not will? The argument is that only God can make water, and He might decide, without any reason, not to, since there is no constraint on God to create or not to create. If we apply normal logic, then His action in this scenario would be random. Therefore, some of the time this chemical formula should not hold, and hence it cannot be a valid scientific formula. Yet this formula $2H_2 + O_2 \rightarrow 2H_2O$ is found to be always true in all tests conducted by the human race[6]. Consequently, we may ask: is it not therefore more logical to conclude [until found otherwise] that God does not interfere in the inner working of the universe, which behaves according to laws of nature, with *a first cause* (which is God)? But Imam Gazzali did not believe in reason or causality. If his anti-causality doctrine was true, then none of the human achievements in science and technology that the world has seen so far, could have taken place. Summarising Gazzali's stance:

- All those who transmit Aristotelian philosophy are *kafirs* (unbelievers)
- The universe is not subject to any natural law
- There is no causality
- God alone is the agent of any event or action in the universe
- Although logic may be studied, mathematics (including arithmetic and geometry) are harmful to the faith
- Physics and other exact sciences should not be studied
- Abstract knowledge and curiosity are dangerous

Imam Gazzali was the most well-known orthodox Muslim genius, who was anti-science and did much to drive reason and rationalism away from Islam. As indicated in chapter 9, the later orthodox scholars were even more anti-rational, and denounced even Gazzali himself as being a suspect. We shall see some further examples of Islamic scholars who fell foul of the orthodoxy in part III of the book.

References and Sources

[Ahm98] Akbar S. Ahmad: Postmodernism and Islam: Predicament and Promise, Routledge, reprinted 1998.
[Arm01] K. Armstrong: Islam – A Short History, Phoenix, 2001.
[Bril71] E. J. Brill (ed): Encyclopaedia of Islam, Leiden. 1971.
[Fak83] Fakhry, Philosophy and History: The Genius of Arab Civilization, J. R. Hayes (Ed), Second edition MIT Press, 1983.

---

[6] As regards the impact of the Uncertainty Principle of Physics on this formula (or on anything else), it is a quantifiable and provable statistical measure (and hence falsifiable), but "If God wills" is not such a measure.

[Hil93] D. R. Hill: Islamic Science and Engineering, Edinburgh University Press, 1993.

[Hoo91] P. Hoodbhoy: Science and Islam, Zed Books, 1991.

[Mar76] Michael E. Marmura: "God and His Creation – Two Medieval Islamic Views", in Islamic Civilisation, R. M. Savoury (ed), Cambridge University Press, 1976. p46-51.

[Nasr78] S. H. Nasr: Introduction to Islamic Cosmological Doctrines, Thames and Hudson (London), 1978, p183.

[Rah79]Fazlur Rahman: Islam, University of Chicago Press, 2nd ed. 1979.

[Rah84] Fazlur Rahman: Islamic Methodology in History, Islamic Research Institute, Pakistan, [1965, 1984].

[Ros67] Ibn Khaldun. Muqadimma, translated by F. Rosenthal, Vol 3, p311, Princeton University Press, 1967 USA.

[Rut91]: Malise Ruthven: Islam in the World, Penguin 1991.

[Sin00] Simon Singh: The Code Book, Fourth Estate (London), 2000.

[Watt63] W. M. Watt: Muslim Intellectual – A Study of Al-Gazzali, Edinburgh University Press, 1963.

CHAPTER 11

FAILURES OF THE LATER ISLAMIC EMPIRES

Three great Muslim empires that dazzled the world in the late middle ages were: Mughal India, Safavid Iran and Ottoman Turkey. Their military strength was unchallenged, their economic prosperity unparalleled and their splendour fabulous. They gained power by military might that was founded on old technology and they later lost power to military might that was created from science-based new technology. In the intervening period there was massive failure to understand and to cultivate science, the real source of power and prosperity. The story of Safavid Iran has been narrated in chapter 2, and therefore we shall focus here on the other two great empires. Mughal power lasted for nearly two hundred years, while Ottoman rule continued for nearly five hundred years. Furthermore, the Mughals ruled over a Muslim-minority region at the edge of the Islamic heartland, while the Ottomans ruled over a Muslim majority region encompassing the Islamic heartland. Therefore there are different lessons for Muslims in the two cases.

11.1 India: The Mughal Empire

Sind in India was conquered by the Arabs in 711 CE, the conquest was later extended into part of Punjab. However the Muslim rule in India began from 1192 under Mohammad Ghuri and it continued up to 1857 when Britain executed sons and grandsons of the last Mughal emperor Bahadur Shah II and exiled him to Burma where he died in 1862. In between the Ghuris and the Mughals, there were three other major dynasties: the Khiljis, the Tughlaks and the Lodhis in Delhi[1], discounting many other minor dynasties in different regions of India, some of whom were independent of Delhi. The area of the empire was largest under the Mughals, which included everything from Afghanistan to Assam and from Kashmir to South India, except the extreme south. Some of earlier independent local Muslim rulers were Shias, but they were swept away by the Sunni Mughals, who nevertheless had a close friendly relationship with the Shia Safavid rulers of Iran and often had influential Shia Queens, chief ministers and courtiers.

Some earlier pre-Mughal emperors contemplated the idea of declaring themselves Khalifas, but gave up that idea for fear of Muslim (and army) revolts. Instead they recognised the Khalifa in Baghdad as the head of the umma, and most of them included the names of both the Khalifa and the Sultan in the Friday Khutbas and on their coins. All of these dynasties were of Turkic origin from central Asia, controlled by the Turkic nobility and of course by the army. However, in contrast to the other Muslim empires, India was a Muslim-minority country, which was ruled by a Muslim king through a relatively liberal policy of toleration, which granted protected status (*dhimmi*) to the majority Hindus (as was granted to the Christians and Jews), without any attempt to convert them. Because of the minority status of the Muslims, the ulama never were as strong as they were in the Abbasid and the Ottoman empires, but nevertheless they had enough clout to create dissatisfaction among the Muslim nobility and the army, unless the emperor was very strong, such as the Mughal emperor Akbar the Great. He was also probably the most liberal of all the Muslim rulers. However, so far as the development of science and technology by the Mughals was concerned, the slate remained fairly clean, as described later in this section.

The most glorious of all the Muslim empires in India was the Mughal Empire, established by Babur in 1526 after defeating the Lodi emperor Ibrahim. Babur was a descendant of Timur Lang on the father's side and Chengis Khan on the mother's side. The term *Mughal* is the Persian word for Mongol, even though the ancestors of Timur probably had some Turkic blood, and furthermore the Mughals used Turkic as their private family language in India until 1760, even though Persian was the court language [Gas91/p15]. Ba-

---

[1] Strictly speaking Delhi was not always the capital, for example, Agra was the capital of the Lodhis and of the early Mughals.

bur was succeeded by his eldest son Humayun (1530-56), who followed his father's advice to treat Hindus and Muslims equally. An opium addict and laidback, Humayun lost his throne to an Afghan warlord Sher Shah in 1540, regaining it, only with the help of the Persians[2], some fifteen years later in 1555, ten years after the death of Sher Shah and just one year before his own death. His wife Hamida Banu, mother of Akbar, was a spirited woman, and his sister Gulbadan Begum [Gul95], friend of Hamida, left an eye-witness account of the reign of Humayun and the early period of the reign of Akbar (1556-1605). Under Akbar the Mughal Empire became the most powerful in the world [discounting China], as decline started in the Ottoman Empire after the death of Sulaiman the Magnificent in 1566 CE [Era04].

Akbar's greatness lies not only in conquering most of India, but also in bringing about Hindu-Muslim unity. He himself married a Hindu Rajput woman, named Jodhbhai, who remained a Hindu and practiced her religion in the Harem. Her brother Raja Man Singh became Akbar's chief general and her son Jahangir was the next emperor. Another Hindu Rajput Toder Mal was Akbar's chief of finance. He also surrounded himself with a collection of wise men, including the poet Amir Khasru, the musician Tansen and the humorist Birbal. However his liberalism attracted the wrath of the ulama, who unsuccessfully tried to depose him. Ultra conservative Sheikh Ahmad Sirhindi (1564-1624), who was a Naqshbandi sufi, issued a fatwa denouncing him as a heretic, but Akbar was too powerful for them.

Akbar was apparently disgusted when he heard about fratricides practiced by the Ottoman rulers, not suspecting that this is what his own descendants would do one day, though not to the same extent. He was succeeded by his son Jahangir (1605-1627), who continued the liberal policy of the father. His eldest son Khasru made repeated attempts to dethrone him, until blinded by the exasperated Jahangir. He also imprisoned Sheikh Sirhindi for his continued opposition to the Mughals, but later not only released him from prison but also became one of his devotees. Jahangir who was gentle and generous

---

[2] To those who are interested in feasts: the Persian emperor Tahmasp I (1524-76) gave Humayun a feast which had mere 1500 dishes (including of course a large variety of fruit drinks), served on gold, silver and porcelain plates, with gold and silver covers! Tahmasp himself outlined in writing the dishes to be served. It would seem that Tahmasp tried unsuccessfully to convert Humayun to Shia Islam.

earlier in life, became sadistic later on, perhaps due to his opium and alcohol addictions. Befitting a Roman emperor, he sent an unarmed man to fight a hungry tiger for fun. He was a naturalist, a good writer and a patron of the arts.

A great influence in his reign was exercised by his wife Nurjahan; his love for her became legendary. She was said to have run the administration from behind the curtains and was so powerful that even her name appeared on coins alongside that of Jahangir. Her brother Asaf Jah (also called Asaf Khan) succeeded their father Itimad ad Dowla, a Shia, as chief minister. She had no children by Jahangir, but had a daughter from her previous marriage. She got this daughter married to Shahriyar, Jahangir's fourth son, with a view to making him the next emperor, thus breaking away from her previous alignment to support Shahjahan, Jahangir's third and ablest son, born of a Rajput Hindu princess called Manmati, who had died a few years after the birth. Shahjahan who married her niece Mumtaz Mahal (daughter of Asaf Jah), outwitted Nurjahan with the help of Asaf Jah who, by order of Shahjahan, captured and killed Shahriyar along with two nephews and two cousins, immediately after Jahangir's death. Nurjahan lost power, but was treated kindly by the new emperor. Earlier she had built the tomb of her father Itimad ad Dowla, which stands three miles downstream from the Tajmahal on the opposite bank of the river Jumna and is second only to the Taj in its elegance. Thus two Persian Shias, Itimad and his granddaughter Mumtaz, had the two most perfect Mughal tombs. Nurjahan also built the simpler tomb of Jahangir in Lahore, where she lived quietly until her death.

Shahjahan (1627-1658) is well-known for building not only the Tajmahal, but also many magnificent mosques, palaces and other structures, including the jewel-encrusted *Peacock throne* (taken away and broken up by Nadir Shah – see below). The famous diamond *Koh-i-nur* (mountain of light) that adorns the crown of the late British Queen Mother was also his. The Tajmahal was built for Mumtaz Mahal, his favourite queen, whose death in 1631 in her 14th childbirth left him grief-stricken for a long time. He was relatively liberal and like his father and grandfather before him he expanded the empire. However, he had a grandiose vision of his own self, claiming to be the *second Timur*, as the 10th descendant of the great conqueror. Shahjahan had four sons and two famous daughters, all from Mumtaz. He designated the eldest son Dara Shikoh the liberal as his successor, ignoring the ablest third son Aurangzeb the orthodox. In a war of succession, Aurangzeb defeated the Imperial army, killed all his brothers and some nephews, and imprisoned (confined) his father to the Agra Fort until the father's death eight years later[3]. His execution of the popular Dara Shikoh and Dara's son Sulaiman on trumped up charges enraged the residents of Delhi, the capital.

Aurangzeb (1658-1707) became the darling of the ulama, reimposed Sharia law and aroused the majority Hindu population against his rule. He spent all his life fighting and died in the south of India, extending the empire furthest, while having a turbulent time even at home. He imprisoned his second son Muazzam on suspicion of embezzlement (the first son died), banished the fourth son Akbar (the ablest) for treachery, and confined his eldest daughter, the highly talented Zibunnesa, a distinguished poet and patron of literature, to the Salimgarh prison for the last 21 years of her life, as a punishment for her secret communications with Akbar. Shahjahan lost out to the children, the children lost out to Aurangzeb. However, the influence of the family of Itimad ad Dowla continued during the reign of Aurangzeb, with his uncle Shayesta Khan (son of Asaf and brother of Mumtaz) as his associate. Personally he was very pious, shunning luxury and living off, it is said, from the income from the sale of the Quran he copied and of some other items, such as religious caps, he himself made,. Given that his general Jaswant Singh was a Hindu, and that he used to fund some Hindu temples, he was probably not as anti-Hindu as it is made out today.

Aurangzeb was succeeded by his 64 year old and weak son Muazzam as Bahadur Shah, but by then the empire had started to disintegrate. The process of disintegration was hastened by the invasion in 1738 of Nadir Shah from Iran who destroyed Delhi, massacring some 30,000 inhabitants. This was followed by the repeated

---

[3] After acquiring the throne, Aurangzeb wanted to visit his father, but protocol required that he the son went to the father at the Agra Fort, where Shahjahan had his eunuch bodyguards. After being warned that if he stepped inside the Fort, he would be killed by his father's eunuchs, he cancelled the planned trip at the last moment, never attempting to visit again. Shahjahan lived in luxury inside the Fort (complete with all his concubines), attended by his favourite youngest daughter Jahanara. Even though Aurangzeb never visited his father, he regularly communicated with him by letters, sometimes seeking his advice on the quality of precious stones, on which Shahjahan was an expert. It is said that Shahjahan died of an aphrodisiac overdose.

destructive raids of Ahmad Shah Abdali of Afghanistan, which weakened Mughal power still further. As with the Abbasid Khalifas, the Mughal emperors were still however regarded as the titular rulers of the whole India, even though they had no power.

In the battle of Plassey in 1757, the Nawab Shiraj ad Dowla of Bengal (which included Bihar and Orissa as well) lost to the English (East India Company), being betrayed by his general Mir Jafar, who succeeded as the subservient Nawab (ruler). From then on the East India Company (EIC) made and unmade Nawabs under the nominal rule of the powerless Mughal emperor in Delhi. The EIC took over the *dewani* (revenue collection) in Bengal in 1765 (giving in return an annual grant to the Mughal emperor Shah Alam II), and in the whole empire in 1803, pensioning off the emperor. By 1852, the emperor's rule was restricted to only the Red Fort (his Palace Complex) in Delhi under EIC control. In the meantime, the disintegration of the empire continued as the local rulers started fighting amongst themselves, with the British (that is, EIC) and the French providing support to the opposing sides. The British side usually won, and after victory they did sometimes take over the kingdom pensioning off the ruler but mostly they made the ruler subservient, as they did earlier with Mir Jafar in Bengal. They soon adopted a new and more aggressive policy of annexing local states as company property by deposing the rulers altogether under one pretext or another. They looted everything they could for the benefit of the Company, and even dismantled many great palaces and mosques in Delhi to sell the marbles for the coffers of the Company. A similar plan for the Tajmahal [Par02/p127] was abandoned after the fall in marble prices[4].

---

[4] In 1831 the Moti Masjid (the Pearl Mosque – made of pearl and marble by Shahjahan) at Agra was dismantled and sold for marble at 125,000 rupees (then £12,500 pounds). At the same time the Tajmahal was priced at 200,000 rupees (then £20,000), but the price was too high for the buyers due to the fall in marble prices. The cost of careful dismantling to save the gems and marbles was considered too high to reduce the sale price of the Taj [Par02/p127]. The other Moti Masjid (in the Red Fort) at Delhi was also saved for the same reason, but was later used as a store-house for munitions. After the Mutiny the Tajmahal was desecrated and used as dance hall for Company employees [Pak02/p192]! The EIC rulers of India were a match for Pizarro (1475-1541), the conquistador, who looted and destroyed, purely for greed, the Inca heritage in South America. Compare the English attitude towards the Indian culture, with that of Charles V (Holy Roman Emperor, 1530-56) towards the Moorish culture in Spain. On seeing the Mosque in Cordoba he immediately halted its further conversion into a cathedral, stating that:

The discontent against the British grew, which resulted in the Indian Mutiny of 1857 CE. Following the Mutiny, the last emperor Bahadur Shah (II) Zafar – an old man, blind and infirm – was exiled to Rangoon (with his two younger sons) where he died at the age of 87 in 1862 in penury. He was a renowned poet and a reluctant symbolic leader of the Mutiny. Not only had he no forces under his command, even his bodyguards were EIC soldiers [Dav04]. With a view to eliminating the line of succession, the British killed his other sons and the grandsons after their surrender [Dal06]. The Red Fort in Delhi with its imperial palaces built by Shahjahan was plundered and much of it was razed to the ground, as tourists can see today. Many great mosques were flattened – the Jami Masjid in Delhi was saved by a hair's breadth [Dal94/pp147-48]. The finest mosques were sold to Hindu bankers to be used as bakeries and stables. After executing thousands of Muslims, the British drove the rest of them out of Delhi, for them to die through starvation [Dal94]. This was the first major application of the famous British *divide and rule policy* in India, given that the Mutiny, which is called the first Indian War of Independence today, was a joint Hindu-Muslim revolt. Surprisingly, Britain was able to persuade the Ottoman Khalifa Abd al Majid I (who needed British assistance) to issue a fatwa urging the 50 million Muslims of India to remain loyal to British rule!

Science, Technology and Medicine

Earlier books on mathematics, geometry, astronomy and medicine were translated during the Muslim rule in India, but most of them before the Mughals arrived and some of them during the early Mughal period. Based on these translations, some new treaties on mathematics (consisting of some applied calculations) were written under the Mughals by Ustad Ahmed Mimari, his three sons and a grandson. The Ustad was one of the architects of the Tajmahal. The

---

"You have built what you or others might have built anywhere, but you have destroyed something unique in the world".
 It may be interesting to compare the sale prices given above with the *annual salaries* paid at that time: The EIC Governor-General of India: £25,000 (higher than the proposed sale price of the Taj), Junior English Army officer in the EIC £200, Indian sepoy in the EIC £10, Indian manual worker £5 to £10, Indian domestics £2 to £3 [DAV04/pp11-35]. The conversion rate at that time was around £1 = 10 Indian rupees.

scientific activities were limited to the translation of a few Sanskrit books on mathematics, some unexciting astronomical observations and the use of calculations for the design of buildings and mechanical devices (for example, to raise water for the Tajmahal) – not a terribly exciting profile.

It seems that some earlier Muslim rulers were interested in, and also developed, some mechanical devices. During the Mughal period, Akbar was keen and participated personally in the development of several mechanical devices, particularly military weapons. But his successors were not so inclined. During the reign of Jahangir, the Imperial army became weak in artillery and modern weaponry. He depended on the Dutch, but did not bother to train his own soldiers to use the artillery or to train Indian craftsmen to make them. In any case the Mughals could not have developed advanced artillery that required mathematical equations for precision, since mathematics (beyond some basic calculations for sharing inheritance) was disapproved.

This dependency on foreign artillery and gunners accelerated under both Shahjahan and Aurangzeb. When the empire started to crumble the local rulers hired artillery and gunners from either the British or the French, who had by then replaced the Dutch and Portuguese as the arms suppliers. Most of the time the British side won, each time the British gaining more power and territory. The Portuguese controlled the Arabian Sea and even the emperor had to pay a toll in Surat for sea-crossings, as was done when Gulbadan (sister of emperor Humayun) went to hajj during Akbar's reign in 1576 CE returning in 1580 [Gul95/pp40-74]. A few ships were constructed during the reign of Akbar, but they were not powerful enough to counter the Portuguese and other sea pirates. For example, a trading ship belonging to the mother of Jahangir was seized by the Portuguese in 1614. Jawaharlal Nehru (the first Indian Prime Minister) criticised Akbar for not having enough foresight to create a powerful navy.

Even though there were royal astronomers, they were more interested in astrology. Some astronomical tables were constructed under the patronage Mughal emperor Muhammad Shah in 1728, directed by Rajput Rajah Jai Singh who built the huge observatory *Jantar Mantar* in Jaipur, that attracts many visitors even today. For them al-Biruni (973-1048, see chapter 6) was the last word in astronomy, the names of Copernicus, Kepler or Galileo were unknown. Sheikh Ahmed Sirhindi, the conservative spiritual leader, used to claim that he could stop the sun from moving! This probably indicates the knowledge of science or scientific rationalism that was prevalent in the Imperial court. Sirhindi unreservedly condemned both philosophy and science, even the study of arithmetic beyond what is required to calculate shares of inheritance and to determine the qibla. Sirhindi did not approve of geometry and trigonometry, let alone algebra or other sciences. He found no merit in knowing that "the sum of the three angles of a triangle is equal to two right angles".

The medicine practiced was known as Yunani (from Ionian meaning Greek), which was based on Abbasid Galenic concepts, often learnt from the translations of the works of, and books on, Ibn Sina, Razi and others (see chapter 8). There were no new techniques or developments, nor any formal schools for teaching medicine, nor any hospital for the sick. Anyone could start practicing medicine. There were no controls on, nor requirements for, any qualifications. For the masses it was the ancient Indian herbal medicine called *Ayurvedic* (Longevity (*ayu*) in the *vedic* way).

Education

Apart from madrasas for religious education, there were no other institutions of education. Mughal princes studied arts, literature, calligraphy, and of course religion, but not mathematics, or science and technology. The following is an interesting commentary on the Mughal education system.

Mulla Shah, who was highly respected by emperor Shahjahan and who taught the royal princes, came to visit his former pupil Aurangzeb after his accession to the throne, expecting to be raised to the rank of *omrah* (high noble). Aurangzeb instead chastised him severely for giving a poor education [Ber99/pp154-161]. Below is a quotation from the long chastisement:

> " .. You taught me that the whole of *Franguistan* (meaning Europe) was no more than some inconsiderable island ... You told me about the king of France and him of Andalusia (Spain) that they resembled our petty Rajas, ... Admirable geographers! Deeply read historians! Was it not incumbent upon my preceptor to make me acquainted with the distinguishing features of every nation of the earth, its resources and strength, its

mode of warfare, manners, religion, form of government, origins of states, their progress and decline, ... errors, changes and revolutions...Far from imparting to me a profound and comprehensive knowledge of the history of the mankind, scarcely did I learn from you the names of my ancestors. ... Forgetting how many important subjects ought to be embraced in the education of a prince, you acted as if it were chiefly necessary that he should possess a great skill in grammar. .. and thus you did waste the precious hours of my youth in the dry unprofitable never-ending task of learning words. ... If you taught me reason, nature of man, sublime conception of the universe, the order and the regular motions of its parts, then I would be more indebted to you than Alexander was to Aristotle. ".

The details of this long speech show that the emperor was very knowledgeable and that he had sound ideas as to what should or should not be in an educational package. Yet he did not see it as his responsibility to devise a proper educational system for his people, instead he left this responsibility to the initiative of individual teachers. When his chief astrologer was drowned by falling into a river, the confidence in astrology also took a plunge, and for a while the court contemplated the idea of introducing science to advance knowledge, as practiced in *Franguistan* (meaning Europe) where astrology was treated as trickery. Nothing however came out of it, and the astrologers soon surfaced from submersion to regain their lost trust!

Economy and Industry

Agriculture was the backbone of the country, all lands belonged to the emperor. The peasants, who were looked down with contempt, had to pay a third to half of their produce as tax to the government, but with relief from taxes during hardship such as droughts and famines. The principal industry was textiles (including silk and carpets), for which India was renowned across the world. Indian muslin was particularly well-known for its fine quality, with incredible stories such as that of 20 meters of the Dhaka muslin being folded inside a single matchbox. However, the artisans were not respected and were paid meagrely. Nevertheless, this industry remained buoyant until

Bengal, the heartland of the production, was taken over by the East India Company, which undertook, it is said, a deliberately destructive policy, with a view to importing English textiles to India instead. Even in 1802, only 15% of textiles were imported from England, but by 1850, the famed Indian textile industry was all but dead, firstly due to the deliberate British policy of strangulation and secondly due to the technologically (and hence relatively cheaply) produced English textiles. One can hear stories even in today's Bengali villages of how the English persecuted the weavers with a view to destroying their industry.

Clearly not only was there no Imperial will, there were no supporting structure and facilities for the development of either science or technology under the Mughals.

Muslim Attitudes

When Sir Sayyed Ahmad Khan (chapter 12) urged Muslims to modernise and to learn English, he was declared a heretic. The Muslim upper class became so out of touch with the needs of the masses that when Persian was replaced by Bengali as the official language in Bengal in 1838, the Muslim upper class fought against it under the slogan "*Islam is in danger*"! With a similar attitude the West Pakistanis tried to force Urdu as the state language in 1952 for the Bengali speaking East Pakistanis, who constituted the majority of the Pakistani population, though occupying a smaller land area. The Pakistani rulers believed Urdu to be a Muslim and Bengali a Hindu language. The East Pakistani Muslims stood up to defend their Bengali culture which the Pakistani military rulers could not tolerate, this resulting in the separation of East Pakistan (as Bangladesh) in 1971 after a cruel and vicious civil war, in which, among many massacres, a generation of Bengali intellectuals were brutally murdered by the Pakistani army and its agents in the name of Islam. The separation proved that the Bengali cultural bond was stronger than the Pakistani religious bond – that is, Bengali Muslims are different from the Pakistani Muslims. If there is any lesson here, it must be that Islam cannot be practiced independently of its local cultural environment – Islam in Jordan will never be the same as that in Japan[5].

---

[5] The Friday Khutbas in the oldest mosque in Japan (Kobe) are usually delivered in English.

11.2 Turkey: The Ottoman Empire

The Ottoman Empire was often referred to in the West as the *gunpowder empire* because of its military might. The dynasty was founded in 1281 by Osman (b 1258, ruled 1281-1324), Uthman in Arabic, whose son Orhan (1324-62) extended the amirate and conquered Bursa in 1326, also making it his capital. His son Murad I (1362-89) was the first great Ottoman sultan, who created a unit of bodyguards, later called *Janissaries* (meaning new armies), for the protection of the sultan. The Janissaries, like the Mamluks (section 2.6), were slaves recruited as boys, converted into Islam and trained for the army, but unlike the Mamluks they never became rulers themselves, although they subsequently played a major role in managing the Ottoman state.

Initially the Ottoman rulers used to call themselves just *sultans* as did many other rulers, but the fourth Ottoman sovereign Bayezid I (1389-1403) after conquering most of Anatolia, wanted to have the more exclusive title *the Sultan of Rum*, formally used by the defeated Seljuk rulers of Anatolia (*Rum* was the Muslim name for the Roman Byzantine Empire). Accordingly he sought and obtained that title from the proper Muslim authority, that is, the Khalifa in Cairo – demonstrating in the process that even the puppet Khalifas had their uses (chapter 2). After extending his amirate to the Balkans, Bayezid suffered defeat at the hands of Timur Lang (1336-1405) in 1402 and committed suicide in Timur's prison in 1404. Timur dismantled Bayezid's amirate, distributing its fragments to the former rulers and also to Bayezid's sons. After a civil war among the sons, Muhammad I (Mehmet in Turkish) succeeded to the throne in 1413 and subsequently conquered back some of the lost territories of his father. He was succeeded by his son Murad II (1421-1451), who recovered most of the lost possessions of grandfather Bayezid I.

Murad II willingly abdicated in favour of his 12 year old son Mohammad II in 1444 and retired to a life of religion, but when his European enemies invaded the empire, he came out of retirement to fend them off. Muhammad II (1451-81) re-ascended to the throne after the death of his father in 1451, conquered Constantinople in 1453 and extended the empire past the Balkans right up to the Danube, thus earning the name Muhammad the Conqueror. In addition to Sharia law which the Ottomans applied rigorously throughout their domain, they also enacted, independent of the ulama (as did the

Abbasids), a set of separate secular administrative laws known as *qanun* for use in administration. Muhammad II enhanced the scope of *qanun*, but for future rulers its relationship with Sharia became a source of conflict with the ulama. Following a civil war between his two sons, Muhammad was succeeded by his younger son Bayezid II (1481-1512), who restored the rights of private property and *waqf* endowments, previously taken away by his father. He also launched a programme to create a naval power and to strengthen the military.

Bayezid II was forced to abdicate in favour of his conservative son Selim I (1512-20), who extended his rule into parts of Shia'ite Iran (but retaken by Shah Abbas of Iran after 1588) and captured Mamluk's Egypt. He brought the puppet Abbasid Khalifa Mutawakkil III from Cairo to Istanbul in 1517, forced him to relinquish the title of Khalifa and then returned him as a private citizen to Cairo where Mutawakkil died in 1543. The title of Khalifa then lapsed. The often-told story, that Mutawakkil III had transferred the title of Khalifa to Selim I before returning to Cairo, appeared first in 1788, and hence is discounted by most historians.

It would seem that all Muslim kings (sultans) after 1258 CE claimed the title *shadow of God on earth* (even in India) and sometimes also Khalifa (not in India). These local Khalifas were meant to be the spiritual leaders of the Muslims in those kingdoms, and not the universal Khalifa of all Muslims. The Turkish sultans also called themselves the shadow of God on earth, but their first recorded use of the title of Khalifa appeared in a letter by sultan Abdul Hamid I in 1774 to Catherine the Great of Russia, in which the sultan claimed spiritual leadership of all Muslims, as against her claim as the leader of all Orthodox Christians and the Pope's claim as the leader of all the Catholics. However, the title of the universal Khalifa – the spiritual leader of all Muslims – was officially claimed in the Turkish Constitution (Article 3) of sultan Abdul Hamid II in 1876. This claim of a pan-Islamic authority, after the loss of much territory to the Christian rulers, was intended to provide Abdul Hamid II with the necessary prestige and authority to hold his crumbling empire together (see later).

Returning to Selim I, he was succeeded by his son Sulaiman the Magnificent (1520-1566), in whose reign the Ottomans reached the peak of their power and their golden age. He was a liberal and he not only created a strong army, but also improved the finances and the administration of justice. For the enforcement of Sharia, he established the post of Sheikh al-Islam (the Grand Mufti) as the third most powerful person (after the sultan and the prime minister) in the land. He expanded the empire from Baghdad to Budapest, incorporating also North Africa and Greece, right up to the river Dniester. Austria was his next target. He reached the gates of Vienna and besieged the city in 1529, but he had to withdraw, vowing to conquer Vienna the next time. When the next time arrived some 150 years later, it was too late (see later). However, during his reign Turkish naval power also reached its height, challenging the all-powerful Portuguese fleet even in the Indian Ocean – a feat that the Mughals were never able to match (see section 11.1).

However, after his death, the army became faction-ridden, and failed to equip itself with new weapons or to learn modern techniques of warfare, as the ulama citing the Quran and hadiths opposed all technical innovations. Sulaiman killed his able eldest son Mustafa on the suspicion of treachery, some claim wrongly. He was eventually succeeded by his drunkard and unworthy son Selim II, whose naval fleet was crushed in 1571 at the battle Lepanto by the Habsburgs, destroying the myth of Ottoman invincibility. As a result the Turkish navy was withdrawn from the Indian Ocean. Selim II was succeeded in 1574 by his son Murad III, when Murad's mother Nurbanu, with the title of Queen Mother, began to control the affairs of the state. The Ottoman Empire reached its greatest extent when the new sultan conquered Cyprus, and then the long slow decline started.

Years later when Murad IV, son of Ahmed I ascended to the throne, all his other brothers were killed following the well-known Ottoman practice, except Ibrahim who was half-mad. Kösem, mother of both Murad IV and Ibrahim became the power behind the throne, like Nurbanu earlier. When Murad IV died in 1640, he had no surviving son to succeed him. Ibrahim was wheeled out from *the special cage* (in which he was kept alive) as the next sultan, and was immediately married off to a woman called Turhan, who promptly produced an heir. When this heir reached the age of seven, he was made the next sultan under the name Muhammad IV in 1648, and his father, the half-mad Ibrahim, was deposed, caged and then executed, sanctioned by an appropriate fatwa from the Sheikh al-Islam, the fatwa pronouncing that there could not be two sultans in the same country! The mastermind behind all these intrigues was the all-powerful Kösem, who did not mind ordering the death of her own son Ibrahim for the sake of the dynasty. But the irony was that when she attempted to dominate Muhammad IV to the exclusion of his mother, Turhan saw red and had Kösem murdered[6] in 1651, after securing a special order from her seven-year old son, the new sultan. Turhan became the new power, but only for five years. This period in the Turkish history is known as *the Sultanate of the Women*, when these two women actually controlled the state apparatus.

These intrigues coincided with internal unrest in the sultanate and the external threat to its security. The long-term enemy Venice was gaining strength and captured some islands from the Ottomans. Muhammad IV (1648-87) tried to bring in technical innovation on the insistence of his son the crown prince, but he was opposed by the ulama. However the prince successfully persuaded the Sheikh al-Islam, who reluctantly gave permission to bring in the printing press for Arabic and other languages, provided that the Quran and other holy religious literatures were not printed. Nevertheless, printing enabled books to be published in Arabic and Turkic that made people aware, not only of the developments in science and technology that were taking place in Europe, but also of how far behind Turkey had fallen (see also later). When Muhammad IV laid the second Ottoman siege to Vienna in 1683, he had to withdraw in disarray after 60 days. The gunpowder empire lost its prestige and power, and began its downhill journey. While the withdrawal of 1529 by Sulaiman was a stalemate for the Ottomans who were still regarded as the strongest power, the withdrawal of 1683 was a catastrophe and was followed by the crushing defeat of the Ottoman army at the hands of the Hapsburgs, with the loss of Hungary and eventually of the whole of central Europe.

With the advent of the 18th century, the Ottomans faced new concerns *vis-à-vis* the rapid rise of military powers in the European countries which threatened their own security. An attempt to West-

---

[6]All the princes (including the child princes) were executed by strangulation with a bow string, so that no royal blood was spilled. Five child princes were killed on the accession of Murad III, at the insistence of his mother Nurbanu and the court. When Murad III died, he left behind twenty very young sons, nineteen of them were murdered by his successor son – an act that shocked the Mughal emperor Akbar when he heard about it many years later. Kösem was given a royal honour, that is, she was garrotted by a bow string, as specified in the execution order signed by the seven-year-old sultan. In India Akbar's descendents did not mind how the brothers were killed, as long as they were dead. Strictly speaking the Mughals were a lot less vicious than their Muslim contemporaries.

ernise was made during the reign of Ahmed II (1691-95), but not very successfully. Sultan Ahmed III (1703-30) sent an envoy to the court of Louis XV with a view to finding out about French civilisation and its achievements in technology (especially military) for application in Turkey. However, the technological decline continued, which concerned Selim III (1789-1807), successor to Abdul Hamid I, who decided to modernise his army. A number of military schools were set up with French instructors to teach mathematics, science and technology to the army cadets. However, taking lessons from the infidels was considered inappropriate by the ulama, who opposed the scheme. Sheikh al-Islam denounced modernisation as the handiwork of the infidels out to destroy Islam, and he issued a fatwa demanding the abdication of the sultan. The ulama conspired with the Janissaries, who first overthrew and then killed the sultan. Another attempt at modernisation was made under Mahmud II (1808-39), who dismantled the Janissaries in 1826, which made the ulama more pliable. The Sheikh al-Islam saw the light and decided to support the sultan. The army was allowed to be modernised and to be trained on new weaponry. A report published by his administration in 1838 recognised the importance of scientific knowledge[Zak89/p123]:

"Religious knowledge serves salvation in the world to come, but science serves the perfection of man in this world".

There were some cosmetic changes as well: the beard gave way to the clean-shave, divans and cushions to chairs and tables, long robes to tunics, and the turban to a compromised fez. These were not enough to catch up with Europe, which was speeding forward fast.

Mahmud II was succeeded by his son Abdul Majid I (1839-61), who launched in 1839 a programme of reform called *Tanzimat* (literally regulations), which included administrative, legal, military and educational reforms. The educational reform was slow and even the plan to establish a secular university did not succeed due to the opposition of the ulama. The sultan produced a constitution guaranteeing the freedom of thought and expression, and removing any distinction between Muslims and non-Muslims – previously unheard of in Islam. The ulama opposed this and incited street riots and dissention. In return for British assistance, the sultan issued a disgraceful fatwa urging the Indian Muslims to accept British rule following the 1857 Mutiny. He was the last Ottoman sultan to die in office.

Due to public disgust, the Ottoman succession by fratricide changed in favour of the ablest family member. Thus Abdul Majid was succeeded by his brother Abdul Aziz (1861-76), who continued the *tanzimat* reforms but his extravagant lifestyle and the need for money to fund the reform programme (to buy arms and equipment) made the country nearly bankrupt, with Britain and France gaining financial control over it. During the period (1868-1874), he also lost vast territories to the Russians, who annihilated his fleet in the Aegean. Liberal reforms, financial fiasco and loss of territories together created a major popular resentment and eventually his dethronement. The next four Ottoman sultans were the sons of his brother Abdul Majid I, except the last Khalifa (not sultan) Abdul Majid III who was his son [Qua00]. Abdul Aziz was deposed in May 1876 in favour of his alcoholic and unfit nephew Murad V (son of Abdul Majid I) and then within a few months, Murad V was deposed in favour of his younger brother Abdul Hamid II in September 1876, all successions legitimised by appropriate fatwas. Abdul Aziz committed suicide within a few days of his dethronement, but Murad V lived for another 30 years. All the while the Ottomans continued to lose territories, mainly to the Russians.

Before his accession to the throne, sultan Abdul Hamid II (1876-1909) pledged to support the new constitution [Lew61/p158, Zur93/pp77-78] prepared by his ministers. Accordingly after his accession, elections were held and a parliamentary government was created, but the conservative sultan ignored the constitution and instead made himself an absolute monarch, returning to strict Sharia and reversing the earlier reforms. One constitutional provision that he adhered to was to declare himself the Khalifa of all Muslims (the universal Khalifa), with the hope of bolstering his regime against the continued territorial haemorrhaging. The point was that as the universal Khalifa he could claim the right to speak up for the Muslims in Russia and elsewhere, paralleling rights already enjoyed by the Russian Czar to speak up for the Orthodox and by the Pope for the Catholics in the Ottoman Empire. He also succeeded in establishing in 1900 the first secular university called *Darul Funun* which was originally planned by Abdul Aziz, and which later became the University of Istanbul.

In 1889 reformist military students in Istanbul established a movement under the name *Committee of Union and Progress* (CUP), also called the *Young Turks* [Zur93/p91], with the objective of modernising Turkey. In 1908 the Young Turks successfully rebelled against the illiberal autocracy of sultan Abdul Hamid II, forcing him to restore the dormant constitution of 1876, and finally dethroning him a year later because of his involvement in a counter-revolution – but the parliamentary government did not improve things. He was succeeded by another brother Muhammad V (1909-18), who entered in the First World War (1914-18) on the side of the Axis powers as demanded by the Young Turks, but after the defeat he fled to Malta in 1918, taking all the blame. Much of the empire was lost in the First World War, but Kemal Ataturk of the Young Turks was able to save Anatolia (mainland Turkey) from subjugation by beating off the Greeks.

In 1918, the National Assembly appointed Muhammad VI (1918-1922), another brother of Muhammad V as the new sultan, but in 1922, at the insistence of Kemal Ataturk, the newly elected National Assembly deposed the sultan declaring Turkey a republic, but retaining the Khelafa with Abdul Majid II, son of the former sultan Abdul Aziz, as the new Khalifa (not sultan). In 1924, Kemal abolished the Khelafa and exiled the members of the Ottoman dynasty. He also made some fatuous changes, such as banning the wearing of the fez and legalising the sale of alcohol and ball-room dancing (see section 12.7).

Paraphrasing Fazlur Rahman [Rah79/p225], secularism in Turkey has been imposed through the militaristic-political power, but has not been embraced whole-heartedly by the people. This perhaps explains the lack of intellectual expressions of secularism in Turkey since the regime of Kemal Ataturk.

Science, Technology and Medicine

The Ottomans, like the Mughals, cared little for the welfare of the people or for culture, even less so for the distant provinces which were given (usually for a fixed term) to the highest bidders for the posts of governors. Once appointed, a governor was free to treat his people anyway he wished. Economic prosperity, particularly of the provinces naturally declined, people were taxed cripplingly, and culture, arts, literature, science and technology withered. The survival of the Ottoman state was founded on two basic principles (i) support for the ulama who legitimise the actions of the state, thus keeping the populace docile and (ii) strong military power (the gunpowder

11: FAILURES OF THE LATER ISLAMIC EMPIRES 183

empire) that can crush any rebellions and can expand the empire as the proof of its prowess. The ulama were made state functionaries, with Sheikh al-Islam at the top – holding the third highest position in the land after the sultan and the grand vizier. It was therefore paramount for the state to keep both sides in tune with each other, which worked well at the beginning. However, when the military strength went into relative decline because of the superiority of Western technology, dissension began to surface due to the opposition of the ulama to the much needed modernisation and technological innovation, which they stifled at every turn. Even though the Ottomans were interested in military technology (and only in military technology), one cannot create and sustain a superior military technology in a vacuum to the exclusion of other technologies and in a society that shuns science.

From the occasional references the Ottomans made to Western science [Lew94/p229], it was clear that they did *not* think in terms of painstaking evidence-based research, transformation of ideas and the gradual growth of knowledge. For them knowledge was something that existed out there, which could be appended to, but could not be nullified, modified or transformed. The scientific concept of evidence and falsifiability was alien to them, even at the highest level. The great scientific enterprise of the Ottoman reign was the foundation of an observatory in around 1575, which appeared to have been destroyed immediately afterwards by the ulama. In 1638, Hezarfen Ahmed Celebi flew over the Bosporus with artificial wings from the Galata Tower in Istanbul[7]. Instead of celebrating this astounding feat, the ulama forced Murad IV to banish Celebi, for the sin of this innovation (*bida'*), to Algeria where he died aged 31. This another example of the attitude of the ulama towards science and technology.

The discoveries of Copernicus, Kepler and Galileo were alien and irrelevant to the Ottomans. These were useless sciences and hence not permissible for the Ottoman ulama. Even in medicine, in which the Muslims had earlier led the world, Turkey fell behind, the sultans hiring European, mainly Jewish, physicians in the court. There were no Ibn Sina, Razi or even Ibn Nafis during the long Ottoman reign. According to Albert Hourani [Hou83/p41]:

---

[7] Hezarfen Airport in Istanbul is named after him – a belated recognition.

"The scientific discoveries aroused no echo [in the Ottoman Empire] – there is no mention of Copernicus in Ottoman literature until the end of the seventeenth century, and then only a fleeting one – and the army and the navy did not adopt the new technical improvements. By the middle of the eighteenth century the evidence of decline was too strong to be ignored."

As regards the military technology, even in 1453 CE, the majority of the gun-founders and gunners were European renegades and adventurers. Turkey also had to rely on the Europeans for the science and technology needed to produce weapons. As stated earlier Turkish naval power reached its peak during the reign of Sulaiman the Magnificent, but it declined fast after the battle of Lepanto in 1571 CE. Over the next 150 years, the control of the seas passed for ever to the Western powers, despite occasional Turkish victories. The supply of European sea rovers to the Ottomans dried up, while the lucrative Atlantic shipping opened up, leaving the Ottomans behind. Technologically backward, the Ottomans had nothing with which to fight back, and they soon became insignificant vis-à-vis the British, Spanish and Dutch naval powers. With the loss of the sea, they also lost trade and wealth. From the late 18th century, the Ottomans, so long self-sufficient, had to place orders for larger ships in the European shipyards. By then they also fell far behind their European counterparts not only in weaponry, but also virtually in all the arts of war, even though they remained far ahead of the rest of the Islamic world [Lew94/p226].

The grip of the ulama was so complete that when a Venetian war galley ran aground in Turkish waters, the Ottoman naval engineers needed an explicit fatwa from the Sheikh al-Islam permitting them to copy some of its features in their own ships, as otherwise they could have been accused of the sin of *bida'* [Lew94/p224]. Innovation was considered to be bad, unless shown to be good by a fatwa. *Bida* was used by the ulama to block technology (including printing – see below) and even medicine. According to Bernard Lewis [Lew97/p289]:

"The decline of the Ottomans was due not so much to internal changes as to their inability to keep pace with the rapid advances of the West in science and technology, in the arts of both war and peace and in government and commerce".

The two most important areas of technological backwardness were shipbuilding and weapons. In addition there were no worthwhile transportation facilities except for pack animals or river and canal boats – there was no proper road network for internal travel, hardly any wheeled vehicles except for a few dignitaries in the city and a few farmer carts in the villages. In the eighteenth century, trade with Europe dropped to 10 percent of what it had been 200 years previously. Traditional Muslim crafts declined, impoverishing the Muslim artisans and craftsmen, reducing them to the level of unskilled labourers. No rejuvenation was attempted by the Sultanate.

The ulama prevented the development of Western technology, as a *bida'*, it was a worse *bida'* since it involved imitating the European infidels. The claims of the modernisers, that the Prophet himself had adopted advanced military techniques of his time from the Zoroastrians and Byzantines and that the earlier Khalifas had also likewise adopted foreign techniques, such as Greek fire, were not enough to justify developments in gun power which required learning from the Europeans [Lew94/p225]. The quotation from the Quran [9:36]: "fight the polytheists completely as they fight you completely", with the reinterpretation that Muslims should employ all weapons, even those of the infidels, did not impress the ulama, keen to avoid the great sin of *bida'*. So the sultans had to hire technology from abroad, but they were not allowed to develop it at home. In any case, it was not possible to keep up with the rapidly advancing European technology without science. As the state lost control of the sea, its income from international trade dwindled, and hence it could no longer even pay for Western arms.

Printing and Clocks

Even the development of the printing technology was obstructed by the ulama as mentioned in chapter 9. Originally invented in China in the 11th century, the Muslims learnt about printing in the 13th century, but this technology was not allowed to take root under Islam by the ulama, on grounds that the use of wooden blocks would defile the "holy" languages (Arabic, Persian and Turkic), let alone God's name, the Quran and the Prophet's hadiths. That printed books of knowledge would become cheap and then fall into the wrong hands also worried some ulama, as noted by a 19th century English traveller E. W. Lane [Sau63] in Egypt. Apparently the ulama wanted to protect Muslims from misunderstanding which might arise, they declared, from the widescale dissemination of knowledge. But according to most experts, the ulama did not want to weaken their own position as the exclusive authority of religious knowledge, by allowing the masses easy access to such knowledge, as would have resulted from the acceptance of printing technology. Note by then the gate of *ijtihad* was closed, and ulama wanted to control all forms of knowledge. However, Christians and Jews were permitted to use printing for their religious books in Hebrew, Greek and other "non-holy" languages. So printing survived, and even flourished in those languages, and eventually spread to Europe in the 15th century.

In the beginning the ban was informal, but when the first printed bible appeared in Germany in 1455, the ulama decided to prevent such blasphemous activities in the land of Islam by making the ban formal. Consequently, a formal edict forbidding the use of printing for the "holy" languages was proclaimed by sultan Bayezid II in 1485, and reinforced by sultan Selim I in 1515. In the meantime, Arabic printing was developed in Italy where an Arabic bible was printed at the beginning of the 16th century. Ambassador Ghiselin de Busbecq of the Holy Roman Empire to the Ottoman court stated about Turkey in a letter dated 1560 [Lew94/p232]:

> "No nation has shown less reluctance to adopt the useful invention of others ….. They have never been able to bring themselves to print books and set up public clocks", on grounds that the holy books would be defiled by printing and the authority of the Muazzin destroyed by public clocks.

As discussed earlier the ban on non-religious Islamic books was lifted in the early 18th century by a fatwa, followed by the lifting of all bans by an Imperial edict in 1727 [Lew97/pp268-9]. However, others claim that the ban was not fully lifted until the early 19th century [Huf94/p225]. In the meantime the Muslims lost the chance to develop print technology, and far worse the ability to disseminate knowledge.

Thus the Muslims who developed a flourishing paper technology in the age of Islamic enlightenment, were prevented from developing the complementary print technology by the ulama in the age of ignorance. Compare this with the Abbasid Golden Age when hundreds of libraries with thousands of books were made available for public access in all corners of the Islamic empire! Where are those books and libraries now? Some books were destroyed by the Mongols, some were declared to contain foreign science and philosophy and hence burnt by the orthodoxy, but there were still others which have disappeared as well. When the curtain of darkness descends, the first casualty is usually the access to books, as in Orwell's 1984, and as happened in Islam. According to T. E. Huff [Huf95/p225-26]:

> "There was in Arabic-Islamic civilisation a strong distrust of the common man and efforts were made after the golden age to prevent his gaining access to printed material".

No wonder then that there was no newspaper in the Islamic world until the mid 19th century. Printing became a thriving business in Syria (which included present-day Lebanon) only after the American Protestant missionaries moved their printing from Malta to Beirut in 1834. The first weekly newspaper appeared in Cairo publicly in 1876, and the first daily in 1889 (called *Muqattam* after a hill outside Cairo) [Huff94/p226]. There was a Muslim Urdu newspaper in Delhi in 1850, but under British rule [Dal06].

We have mentioned above the ban on clocks as reported in the letter of Ambassador Ghiselin de Busbecq in 1560. As with printing the ban on clocks was later revoked, again when it was too late to compete. Sundials and water clocks were known to, and had been improved upon by, the earlier Muslims, but not the mechanical clocks and watches which were invented in Europe in the fourteenth century. In the sixteenth century, the Ottomans used mechanical clocks and watches that were imported from Europe. From that time mosques were also permitted to have clocks. In the seventeenth century, some European émigrés (non-Muslims) even produced excellent watches and clocks in Istanbul following some Swiss and English designs, but the craft died down when these devices became too sophisticated for the Ottoman technology and when the West placed an embargo on the export of parts.

Finally, Greek mathematics declined due to lack of interest in it by the Romans, whose building and engineering works required only low-level arithmetic. At that time the Romans had no competitors and no country was doing any better than them in technology. This argument is difficult to sustain in the later Islamic period when the Mughals, Safavids and Ottomans, particularly the Ottomans, were aware of the superior Western technology based on new science, but were unable to develop it mainly due to the objections of the orthodoxy.

References and Sources

[Ber99] Francois Bernier: Travels in the Mughal Empire (AD 1656-1668), translated from French by Vincent A. Smith, 2nd edition, Low Price Publications, Delhi, reprinted 1999, pp154-61.

[Bos76] Edmond Bosworth: Armies of the Prophet, Islam in the Arab World, (Ed: B. Lewis), A. Knopf, NY, 1976, pp 201-24.

[Dar94] W. Dalrymple: The City of Djinns, Flamingo, p147-48, 1994 – an excellent informative history of Delhi ruins.

[Dal06] W. Dalrymple: The Last Mughal – The Fall of a Dynasty, Delhi 1857; Bloomsbury, 2006. A great book to read.

[Dav04] Saul David: The Indian Mutiny 1857, Penguin, 2004. Very informative, though written from a pro-British stance.

[Era04] Abraham Eraly: The Mughal Throne, Phoenix, 2004 – an excellent book.

[Gar11] Lucy Garnett: Turkey of the Ottomans, Pitman, 1911.

[Gas91] B. Gascoigne: The Great Mughals, Jonathan Cape, reprinted 1991 – an excellent British TV series was based on it.

[Gof02] Daniel Goffman: The Ottoman Empire and the Early Modern Europe, Cambridge University Press, 2002.

[Goo99] Jason Goodwin: Lords of the Horizons, Vintage, 1999.

[Gul95] A. S. Beverage: History of Humayun by Gulbadan Begum, Low Price Edition, Delhi, 3rd reprint 1995. Gulbadan was the sister of Humayun, a close friend of Hamida Banu (Akbar's mother). The book is a real eye opener of the then Imperial life.

[Hab79] Irfan Habib: Changes in Tech in Medieval India, in Technology and Society, Indian History Congress, Waltair, 1979.

[Hou83] Albert Hourani: Arabic Thought in the Liberal Age (1798-1939), Cambridge University Press, 1983.

[Huf95] T. E. Huff: The Rise of Early Modern Science, Chapter 2, Cambridge University Press, pbk edition 1995.

[Itz76] N. Itzkovitz: The Ottoman Empire, Islam in the Arab World, (Ed: B. Lewis), Publisher: A. Knopf, NY, 1976, pp273-300.

[Lew61] B. Lewis: Emergence of Modern Turkey, 1961.

[Lew73] Bernard Lewis: Islam in History, Alcove Press, London, 1973.

[Lew94] B. Lewis: The Muslim Discovery of Europe, Phoenix, 1994.

[Lew97] B. Lewis: The Middle East, 3rd edition, Phoenix Giant, 1997.

[Lew03] B. Lewis: The Crises in Islam, Weidenfield and Nicholson, 2003, p88/89.

[Maj46] R. C. Majumdar, et al: Advanced History of India, Macmillans, London 1946 – an excellent scholarly book.

[Par02]: Fanny Parkes: Begums, Thugs and White Mughals, Sickle Moon Books, London 2002, edited by W. Dalrymple from the Journals of Mrs Parkes who kept this diary as she lived (with her husband – a Company employee) and travelled in India in the eighteenth century before the Mutiny. A very interesting book.

[Par76] V. J. Parry: A History of the Ottoman Empire to 1730, Cambridge University Press, 1976.

[Pei94] Leslie Pierce: Imperial Harem, Oxford University Press 1994 – a hard to read but scholarly book.

[Qua00] Donald Quataert: The Ottoman Empire, 1700-1922, Cambridge University Press, 2000. The book gives a genealogy of the Ottoman sultans.

[Rah79] Fazlur Rahman: Islam, University of Chicago Press, 2nd ed. 1979.

[Riz76] S.A. Rizvi: Muslims in India, Islam in the Arab World, ed. B. Lewis, published by Alfred A. Knopf, NY, 1976, pp310-320.

[Rut91]: Malise Ruthven: Islam in the World, Penguin 1991.

[Sau63] J. J. Saunders: "The Problem of Islamic Decadence", Journal of World History, Vol 7, 1963, pp701-720.

[Sau65] J. J. Saunders: A Medieval History of Islam, Routledge, 1965.

[Say86] Sayili, Aydin: Turkish Contributions to Scientific Work in Islam, 1986.

[Zak89] R. Zakaria, The Struggle Within Islam, Penguin, 1989.

[Zür93] Erik Zürcher:Turkey – A Modern History, I. B. Tauris, 1993. (see pp76/77 on the constitution of 1876).

PART IV

FUTURE PROSPECT

CHAPTER 12

SOME LATER REFORMERS

As the three great Muslim empires started their decline during the eighteenth century, the technologically superior European powers began to make inroads into their territories. In reaction to the Western colonisation, many resistance movements grew in many Muslim countries in both Asia and Africa. Most of the resistance leaders believed that once liberated from Western domination, they would be able to revitalise Islam by returning to the time of the Rashidun and recreate the old glory. They never realised that the world had changed, the past solutions could not be successfully applied to the new complex problems of the technology driven world. The leaders that fall into this category include the Khalid Baghdadi (1767-1827) of the Naqshbandi sufi movement (disciple of Sirhindi of India), Baha al-Din Vaishi (1804-93) in Kazan (Russia), Yaqub Beg (1820-77) in Turkistan, Shah Wali Allah (1702-62) in India, Haji Shariat Allah (1781-1840) in Bengal (India), Muhammad Ahmad [Mahdi] (1840-85) in Sudan, Muhammad Ali al-Sanusi (1787-1859) in Libya, Ahmad bin Idris (1760-1837) in Morocco and Muhammad Abduh (1849-1905) in Egypt. There were many others besides.

However, there were some others who viewed modernisation as the key. Some scholars, such as Jamaluddin Afghani, saw clearly the need for a science-based society, but were less concerned by the deep conflict between science and orthodox Islam. They did not appreciate that a modern society could not be created without some fundamental changes in the orthodox religious attitude. Nevertheless, there were a few – a very few – reformists who saw the need for a fundamental change in the nature of orthodox Islam to develop a science-based modern society. Of these reformists two names shine above others: Sir Sayyed Ahmad Khan and Syed Ameer Ali, both from India. These two scholars, along with Afghani, are perhaps the most pro-science reformers of Islam, unsurpassed by anyone else even today, and therefore we shall present their views in this chapter.

In addition to these reformers, there were also some rulers who were keen on modernisation, which they tried to impose from the top, ignoring the problems of orthodox Islam in a modern society. Since modernisation is the key to the development of science and technology, it is important for us to review these efforts in order to understand the reasons for the failure to modernise. The rulers of our interest are: Muhammad Ali of Egypt, King Amanullah of Afghanistan, Reza Khan of Iran, and above all Kemal Ataturk of Turkey. As we review their works below, we shall consider the scholars first and the rulers second.

12.1 Sir Sayyed Ahmad Khan (1817-1898)

Sir Sayyed Ahmad Khan, who came from an aristocratic family, was far ahead of his contemporaries in his grasp of the problems faced by Muslims in the post-Mughal India. When the new English-based educational programme was launched by Macaulay around 1829, it was welcomed by Hindus with open arms, but boycotted by Muslims who declared the study of English as *haram* (forbidden).

Sir Sayyed learned from the trauma of the failure of the Sepoy Mutiny of 1857, and became convinced that desperate remedies were needed if the Muslims of India were to be anything but "stable boys, cooks, servants, hewers of wood and drawers of water". He saw the backwardness of Muslims as a direct result of superstitious beliefs, rejection of reason and blind obedience to tradition, which he believed could only be countered by the reinterpretation of Muslim theology, to make it consistent with the Western humanistic and scientific ideas, extracting in the process "pure" Islam from the "fossilised" and irrelevant dogma.

He advocated a reinterpretation of the Quran in order to remove all contradictions with scientific findings. Since the Quran is the word of God, it cannot be untrue and therefore any contradiction with scientific knowledge could only be apparent, not real. He suggested the following methodology for the reinterpretation of the Quran:

1. A close enquiry be made into the use, meaning and etymology of the Quranic language, so as to yield the true meaning of the words and passages.
2. The criteria employed to decide whether a given passage needed metaphorical interpretation, and which of several interpretations ought to be selected, should be founded on the higher truth established by science. Such truth is arrived at by *aqil dalil* (rational proof) and demands firm belief.
3. If the apparent meaning of scripture conflicts with demonstrable conclusions, then it must be interpreted metaphorically.

The third point is supported by the Quranic verse [3:07] as discussed in section 3.1. In fact the whole approach of Sir Sayyed on the interpretation of the Quran is similar to that of the Mutazilites (chapter 3) and also conforms to that of Ibn Rushd (chapter 10).

Clearly if scripture is reinterpreted and reason is employed, then the present basis of Sharia law devoid of human condition and reason (chapter 4) would crumble, and much of it would become untenable. Consistent with this view, Sir Sayyed dismissed Sharia law as irrelevant to modern times. He was aware of the wrath of the orthodoxy he would encounter, and stated:

"My enquiring mind ... made me arrive at the truth which I believe to be *thet* Islam (pure Islam) although conventional Muslims may hold it to be *thet* kufr (pure unbelief)".

He was a pro-imperialist untouched by the liberation theory, one who saw the Indian Muslims as loyal British subjects, making progress under the "benign" British rule – a view that did not make him popular. In fact many fatwas were issued denouncing him as a *kafir* (heretic), not only for his promotion of English, but also for his other reformist views. The *mutawalli* (keeper) of the Kaaba declared him an enemy of Islam and *wajib-i-katl* (deserver of death).

However, his greatest legacy was the establishment of the Muhammadan Anglo-Oriental College in Aligarh in 1877, which became the Aligarh Muslim University in 1920, now in secular India the University of Aligarh, where a generation of upper class Muslims were educated during earlier times. However, all of them were upper-class Muslims, who dominated the Indian Muslim League, which gave birth to the state of Pakistan.

12.2 Syed Ameer Ali (1849-1924)

Educated in England, Syed Ameer Ali was a reformist in the mould of Sir Sayyed Ahmad. The two great books that he wrote are (i) Legacy of Islam and (ii) Spirit of Islam. Both books are still widely referenced by scholars, but the second book is considered to be his greatest contribution to liberal Islam, first published in 1891, and updated continually up until 1922. Ameer Ali saw the Prophet as a "great Teacher", as a believer in progress, as an upholder of the use of reason, and also as "the great Pioneer of Rationalism". He regarded Islam as an ideal religion, a religion that not only insists on a true belief in God, but also encourages learning and science, and emphasises human responsibility and freewill. In his book, he presented evidence in favour of these assertions and argued against contrary views. This book was translated into many other Muslim languages, but the orthodox responded to this great work by dubbing its author as a Western apologist – a danger to Islam.

The summary of his views on knowledge and science may be given as:

- The Quran and hadiths regard knowledge (which includes science) as a supreme attribute that motivated the early Muslims to study science to understand God's greatness
- There is no conflict between Islam and rationalism. The Mutazilites, apart from some excesses, made a major contribution to knowledge. Those Muslim philosophers and scholars, from Kindi to Ibn Rushd, were our heroes.
- The fanatics and rigid dogmatists caused Islamic science and innovation to collapse, the ones that bear the most responsibility for this collapse include Ashari, Ibn Hanbal, Gazzali and Ibn Taymiyya.
- Science needs to be brought back to Islam

He wrote: "For five centuries, Islam assisted in the free intellectual development of humanity, but a reactionary movement then set in, and all at once the whole stream of human thought was altered. The cultivators of science and philosophy were pronounced beyond the pale of Islam". In his opinion Islam needed to be rescued from the Imams and Mujaddids (orthodox reformers), and the minds of Muslims freed from the bondage of literal interpretations. He likened Mutazilites fighting against the Sunni orthodoxy to Protestantism fighting against Roman Catholicism.

Both Sir Sayyed and Ameer Ali advocated liberalism in Islamic society, including the abolition of polygamy and purdah. They regarded jihad as the intellectual struggle.

12.3 Jamaluddin Afghani (1838-1897)

Jamaluddin Afghani was a major Muslim leader of the 19th century, who influenced the Muslim world in its struggle against Western colonialism and was known as the Sage of the East in Arabic literature. He was also a pro-science modernist reformer.

He was born not in Afghanistan as he himself claimed, but in a Shia family near Hamadan in Iran, although he later became a Sunni. He studied the thoughts of Muslim rationalists in early life and was particularly influenced by Ibn Sina, which led him to develop a pro-science and rationalistic attitude. He spent his youth in India during the period of Sepoy Mutiny of 1857 when he concluded British imperialism to be the greatest evil facing Muslims. Afghani was an orator, a journalist and a political activist. He saw *ijtihad* replacing *taqlid* as the key to reform. He believed religion to be rational like science without any contradiction between them. He lived and preached his message also in Afghanistan, Iran, Turkey, Egypt, Russia, France and England, but was expelled from several Muslim countries for his unacceptable political activities. He travelled to Turkey in 1869, befriended the Ottoman sultan Abdul Aziz (1861-76) and buried himself in educational reforms. He was later declared a heretic and expelled from Istanbul for comparing the obstructionist Sunni ulama unfavourably against the men of science, in connection with his attempt to create a science university (which was eventually established by Abdul Hamid II in 1900 as *Darul Funun* – chapter 11). The fact that he had a Shia background might have compounded his problem.

Undaunted Afghani entered Egypt in 1871, lectured on his reformist thoughts to young recruits such as Muhammad Abduh (a future reformist leader) but was expelled in 1879 for anti-*Khedive* activities. He went back to India to resume his reformist preaching.

Advocating the cause of science he stated in a lecture in Calcutta in 1882:

> "There was, is and will be no ruler in the world except science ... the benefits of science are immeasurable; and these finite thoughts cannot encompass what is infinite."

In the same speech, he lamented the state of Muslim scholarship and their rejection of philosophy, logic, science and literature, particularly deriding the Indian ulama for their ignorance, especially of science. He continued:

> "... they do not ask: Who are we and what is right and proper for us? They never ask the causes of electricity, the steamboat and railroads ... our ulama at this time are like a very narrow wick on the top of which is a very small flame that neither lights its surroundings nor gives lights to others."

He further wrote: "The human society must be freed of all supernatural belief if it wishes to work on its essential work which is the construction of positive science. This does not imply violent destruction nor brusque ruptures". He criticised Muslims for not separating the spiritual and temporal domain and he thought the orthodox had succeeded in using Islam to stifle science and to stop its progress. "I cannot keep from hoping" he said "that Muhammadan society will succeed someday in breaking its bonds and march resolutely in the path of civilisation after the manner of Western society".

On the futile search for knowledge by an orthodox believer, he said:

> "A true belief [according to the orthodox] must in fact turn from the path of studies that have their object scientific truth ... Yoked, like an ox to the plough, to the dogma whose slave he is, he must walk eternally in the furrows that have been traced for him in advance by the interpreters of the law. Convinced besides that his religion contains in itself all morality and all sciences, he attaches himself resolutely to it and makes no effort to go beyond ... What would be the benefit of seeking truth when he believes he possesses it all? ... Whereupon he despises science."

For all his rhetorics, unlike Sir Sayyed Ahmad Khan and Ameer Ali, he did not examine deeply the inherent conflict between the religious orthodoxy and scientific rationalism, which requires resolution through the reinterpretation of scripture, as was attempted by the Mutazilites (chapter 3). In contrast to Sir Sayyed, Afghani was an anti-imperialist religious reformer, who wanted liberation from foreign rule and an ideal Muslim state, in which science was expected to flourish. He cared less for science and technology, which was secondary to him – something that could come later after the liberation. Afghani saw Sir Sayyed Ahmad Khan, as did many others, as despicable – an agent of the British out to destroy Islam.

After visiting many other countries, in 1892 he returned to Turkey, where he preached pan-Islamism (which he had developed earlier with Muhammad Abduh in Egypt), while continuing to foment anti-British anti-Shah revolt in Iran. He however disgraced himself and his cause in 1896 by arranging and ordering the assassination of the pro-British Shah of Iran, Nasir al-Din. He was able to escape the gallows because of his friendship with the Ottoman sultan Abdul Hamid II (1876-1909) who refused to extradite him to Iran. He died in Turkey in 1897. Nearly half a century later, the government of Afghanistan claimed him as their citizen and brought his remains from Istanbul to Afghanistan for re-burial in a shrine at Kabul University in 1944.

However, the liberal utterances of these reformist scholars did not make much impact on Indian Muslims, which Sir Hamilton Gibb in 1945 attributed to "the failure to grasp the fact that no endeavour can succeed unless it achieves a balance between the broad and the deep currents of a people's psychology, and the inescapable forces of social evolution". Montgomery Watt interprets it as the conflict "between a people's deep attachment to their self-image and the consequences of being part of the twentieth century world" [Wat88/p 64]. What Watt alluded to is the Muslim self-image that all knowledge can be found within Islam (*Closed World Assumption* in knowledge) and that it is unnecessary (and even *bida'*) to seek any knowledge from outside Islam (*Open World Assumption* in knowledge), a point that concerned Afghani as well.

12.4 Muhammad Ali of Egypt (1769-1849)

Following Napoleon's invasion of Egypt in 1798 CE to curb British influence, the Ottomans sent an army of Janissaries and Albanians, under the command of an Albanian officer called Muhammad Ali, to drive him out with help of the British. In 1801, the French were thrown out, but the Mamluks (chapter 2) refused to accept the new governor (pasha) sent by the Ottoman sultan to rule Egypt. Taking advantage of this unrest and with popular support, Muhammad Ali the army officer seized control of Egypt, forcing the weak sultan to accept him as the new pasha of Egypt. From then on, his descendants ruled Egypt up until 1952, when they were overthrown by an army coup.

Muhammad Ali was an ambitious man who wanted to make Egypt a modern state independent of the Ottomans. He was impressed by the French technology he saw during his fighting and he wanted to emulate it in Egypt. He was an uneducated man of peasant stock, who learnt to read only in his forties. He had no interest in the intellectual revolution that had taken place in the West, all he wanted was to acquire modern technology and military might, unwisely ignoring the attendant consequences of such modernisation on the religious, cultural and political life of Egypt. Nevertheless, his achievement was remarkable, since by the time he died in 1849, he had almost single-handedly dragged Egypt from its existence as a backward Ottoman province to a forward-looking country in the modern world, but at a heavy price.

He undertook the most complex transformation of a country that had been left in an appalling state by the Mamluks. The evidence of their pillaging and destruction was everywhere, even fellahin (peasants) had deserted their lands and migrated to Syria, due to the crippling taxes of the Mamluks. On top of all these, like a black cloud on the horizon, the Mamluks still posed a major threat to Muhammad Ali's power. He decided his first task was to physically eliminate them. In 1805, he enticed the Mamluk beys (lords) to a meeting, and there he killed most of them, the remainder were killed over the next few years, through treachery and conspiracy. Some 1000 other Mamluks were massacred in one day, their houses looted and women raped. At long last, after 500 years, Egypt had become free of the Mamluks.

12: SOME LATER REFORMERS

His next task was the Egyptian economy. Between 1805 and 1814, he made himself the personal owner of every piece of land in Egypt. Having already appropriated all the Mamluk estates, he then seized land from the farmers, and finally he confiscated all the *waqf* lands donated to the religious institutions and other good causes. With similar ruthlessness, he made himself the owner of all trade and industry in Egypt. Thus within ten years, he became the sole landlord, merchant and industrialist in Egypt. Egyptians went along with this grabbing, not only because of the generous compensation paid to them, but also because Muhammad Ali used the proceeds, not to line up his own pocket, but to develop Egypt. He was able to establish law and order, a fair justice system and proper management of the government after years of chaos, misrule and mismanagement. Anyone with a complaint could approach Muhammad Ali directly. His greatest achievement was the cultivation of cotton, which was exported to make Egypt rich and to give him funds to import European technology for Egypt.

He was determined to build a strong industrial base, which he knew to be essential for the future of Egypt. As he did not understand the intellectual, social and scientific requirements and the consequences of innovation and modernisation, he opted for imitation. He just went ahead, without thinking, with the establishment of a sugar refinery, an arsenal, copper mines, cotton mills, iron foundries, dyeing works, glass factories and printing works. But industrialisation takes time, and for a start one needs a trained workforce. The fellahin recruited to work in these industries were uneducated, had no technical expertise, no experience, no sense of productivity and were unable to adapt to factory life away from their fields. Consequently most of Muhammad Ali's industrialisation was doomed to failure and it did fail.

He also introduced Western style administration, manned mostly by Europeans, Turkish and Levantine officials. Bright young Egyptians were sent to Europe for further studies. A military college and some artillery schools were established, with European and Western educated Egyptians tutors, to teach science among other things. Thus the pasha (i.e. Muhammad Ali), who was in a hurry, got trained army officers and an educated upper class, but he made no provisions for even the primary education of the fellahin, who were dispossessed and excluded. This did not help the creation of a fair society nor the creation of industries. As his chief concern was the army, this social and educational gap did not strike him as a major problem in development and industrialisation. However, he was able to keep Egypt out of British and Ottoman control, partly by accepting the nominal sovereignty of the Ottomans and by helping them in their needs with his powerful fighting machine.

He recruited 20,000 Sudanese as conscripts to fight in his modern army, but most of them died in their camp in Aswan, as they could not cope with the rigour of a modern army. He replaced them by Egyptian fellahin as conscripts, forcing them away from their homes, families and fields. As the conscripted fellahin were the bread winners, their families became destitute and their women folk prostitutes. Their situation was so desperate that many men resorted to self-mutilation to avoid conscription. As a result, not only the fellahin but also agriculture as a whole suffered. Worse was to follow. Ironically his economic growth left Egypt open to the European trade, which resulted in the bankruptcy of local trade and industry, as they could not compete with the West. As he was the sole merchant and industrialist, the indigenous merchant and entrepreneurial classes were destroyed. He invested a vast amount of money in irrigation and water communication works, but the working conditions were so poor that 23,000 people are said to have perished in these ventures.

There was an ancient tradition of conscription in the Egyptian army. Even the Pharaohs used to conscript Sudanese (Nubians) and Egyptian fellahin, but in those ancient days the dislocations were deliberately kept small and dispersed; care was taken not to disrupt the agriculture, the backbone of Egypt. The people were recruited for a shorter time, and the condition in the army was not entirely alien. But Muhammad Ali, an action man, was instinctively opposed to thinking, never reflecting on the consequential social issues; perhaps he was not intellectually capable of doing so.

Two Egyptian societies grew, the rich who had everything and the poor who had nothing. The rich included the military and administrative personnel, who were modern and liberal, but the vast majority were the poor, conservative and very backward. The ulama saw the modernisation as a destructive enterprise, even though they supported Muhammad Ali initially for freeing them from the Mamluks. He frightened them with financial starvation (as he owned all the waqf properties), and pacified them with lip services to Islam. He deposed any sheikh who defied him, fully knowing that such a deposed sheikh could not last without an income from him. Many mosques and madrasas fell into ruins, even al-Azhar reached a wretched state. The ulama were left behind, instead of being recruited to develop a theology that encouraged modernisation and could thus generate mass support for the reform.

The crucial question is: why did his modernisation not bring Egypt up to the European level? The obvious answer is that a crash modernisation does not work. It requires intellectual, social, religious and political evolution, and it needs not only capital but also an educated population. These things have to evolve, a ruler can expedite this evolution by providing the necessary facilities and by creating the necessary infrastructure, but he cannot impose it on an unwilling or unprepared society. It would also appear that the left-behind ulama in Egypt viewed modernisation as innovation (*bida'*), and hence unacceptable. As the European reformation has shown, modernisation is not possible unless the society encourages free-thinking, and welcomes technological progress and innovation.

In 1867, his grandson Ismail (1830-1895, ruled 1863-79, deposed) bought the Persian title of *Khedive* (great prince) from the Ottomans, at an extra annual tribute of £350,000 to distinguish him from other Ottoman pashas, but he became unwisely very pro-British, turning to them for help and protection. This was the time of the construction of the Suez Canal. The anti-British feelings heightened in Egypt after the Anglo-French acquisition of the Egyptian shares in the Canal through an unfair business deal – a deal that was substantially responsible for ruining the Egyptian economy and for bringing the country close to bankruptcy. Egypt launched a rapid modernisation programme, but without having the necessary institutions or the required understanding to support, guide and sustain it. As Egypt could not pay for this programme after the disastrous Canal deal and after a sudden sharp fall in cotton prices (following end of the American Civil War), it fell victim to the greed of European banks and their cohorts. The already precarious economic position of Egypt worsened, accentuated by unscrupulous Western business deals, but the country was kept on a life-support machine by the Western powers, who took over the control of the Canal in order to preserve their self-interest, ruining Egypt further.

Subsequent unrest against the misrule of the Khedive and British influence was crushed with the help of British forces, resulting in further British domination. In 1883, Sir Evelyn Baring (later Lord

Cromer) arrived in Cairo, ostensibly to act as the British Consul General, but actually to rule Egypt, even though the country was legally part of the Ottoman Empire and remained so until 1914. At the outbreak of the First World War in 1914, Britain declared Egypt its protectorate violating Turkish suzerainty and deposed the pro-Ottoman Khedive Abbas II in favour of his uncle Hussain Kamil, upgrading at the same time the title Khedive to *Sultan* [Vat91]. With the connivance of the pro-British Egyptian Prime Minister Rushed Pasha, Britain bled Egypt for the war effort. It even conscripted 1.5 million fellahin into Labour Corps, forcing them out of their families and lands into the various Middle Eastern war zones. The peasants were ruined, but a few big landowners made money from overpriced cotton. The British rule led to untold misery and anti-British resentments. When Hussain Kamil died of ill-health in 1917, his only son Kamal al-Din Hussain declined to be the next sultan in this situation. Britain then installed Ahmad Fuad, the brother Hussain Kamil, as the sultan. In 1922, after much unrest against it, Britain decided to give independence to Egypt under sultan, now king, Fuad. But Britain continued to exert influence and manipulate king Fuad and his successor Faruq, against political opposition in Egypt. Faruq was deposed by Nasser's coup in 1952.

12.5 King Amanullah Khan of Afghanistan (1892-1960)

King Amanullah Khan of Afghanistan came to the throne after ousting his own father. He declared and won the third Afghan-British War in 1919, forcing the British to recognise the independence of his country. Emboldened by the enhanced prestige, he embarked on modernisation between 1923 and 1924, including the creation of a Parliament (he chose the Members) and granting freedom for women. His beautiful young wife Surayya tore up her veil dramatically in public in support of women's freedom, to the utter horror of the ill-educated conservative Afghans.

As expected, his reforms infuriated the mullahs and the tribal chiefs alike. Very few Afghans had seen Western dress at that time, let alone had the experience of wearing it. And yet, in the name of modernisation – perhaps following Ataturk (section 12.7) – the King demanded his Members of the Parliament to wear Western dress. The Members obeyed their King and arrived in the Parliament with ill-fitted, often torn and shabby Western clothes, some even wearing clothes inside out. That was the spectacle of modernisation.

The King went abroad on a foreign tour, but pictures of his wife participating in innocent ball-room dancing were circulated back to the tribal chiefs in Afghanistan. Apparently this circulation was organised by the British agents bent on destabilising the kingdom, in revenge for their humiliating defeat in the third Afghan war, the only war in the 20th century that Britain lost. The ensuing tribal rebellion spear-headed by the mullahs and fomented by the British, forced the anti-British Amanullah to abdicate in 1929 in favour of his *elder* brother Inayetullah, thus ending the reform experiment. He went into exile in Switzerland and died in 1960.

The ineffective Inayetullah Khan was overthrown soon afterwards, with the help of some tribal chiefs, by a Tajik high-way robber called Bachha-i-Sakao (Child of Sakao) who declared himself King Habibullah Ghazi. Amanullah's general Nadir Khan crushed the Bachha and became King Nadir Shah. He was assassinated in 1933 and was succeeded by his 19-year old son Zahir Shah who was relatively liberal but was overthrown in 1973. Then arrived the Soviet communists promising reform but had to withdraw in defeat and disgrace. The resulting vacuum was filled in by the Talibans, followers of an ultra-conservative (Saudi-backed) Wahhabi doctrine, who reduced the country into a draconian primitive state. The Talibans were ousted in 2002 by an International force sanctioned by the UN and led by the USA. Zahir Shah, then 85, returned to Kabul from exile as the former King (to live as a private citizen), but he died in 2007, aged 92. The turbulence in Afghanistan continues.

12.6 Reza Khan [Shah] of Iran (1877-1944)

Weak and incompetent, the Qajar rulers gave up the idea of modernising the army after only a half-hearted attempt. This suited both Britain and Russia admirably, since they favoured a weak independent, but backward, Iran as a buffer state between British India and Russia. They successfully blocked the modernisation of Iran as it could have been detrimental to their strategic interests. For example, towards the end of the 19th century, both the powerful neighbours (i.e. British Empire and Russia) extracted heavy economic concessions, such as railway construction, all mineral extractions, irrigation networks, a national bank and various industrial projects, but they deliberately embarked on a programme that furthered only their own interest (such as mineral extraction), while blocking those reforms (such as a railway system) that could have benefited the Iranians. Britain also signed a contract to modernise the Iranian army with British military equipment, without any intention of fulfilling it. But that was not all. As Iran was too fearful to take strong action against either of these two powers, they started interfering with the Iranian administration, with blatant exercise of political control. This control, together with the penetration of British and Russian merchants and their goods in favoured terms, decimated Iranian merchants and industries.

The only people who were not afraid to oppose were the conservative Shia ulama. Unlike Muhammad Ali of Egypt, the Shah did not have a strong army nor a central bureaucracy to impose his will on the ulama, who lived off a secure financial income from waqfs trusts. Supported by the merchants, the ulama began a campaign against the Shah. Hitherto non-political, the Shia ulama became politicised as an anti-foreigner (and also anti-modernisation) force. They objected even to modern military schools, staffed by undesirable foreigners, for teaching undesirable science and technology (including undesirable mathematics). To them the schools were European ploys to destroy Islam. The opposition of the ulama, made the Shah even more dependent on the Western powers for his political survival, resulting in further concessions being granted to Britain, which established in 1909 the Anglo-Persian Oil Company, the forerunner of the British Petroleum. The Britons in Iran were beyond the Persian laws, rarely ever punished even for gross offences. Britain started to rig Iranian elections to the Majlis (Iranian Parliament).

In 1921, Reza Khan, the Commander of Shah's Cossack Brigade overthrew the Government, became prime minister, and started courting Russia and the USA in order to marginalise Britain. In 1925, he forced the Qajar Shah to abdicate with a view to creating a republic. But the ulama objected to the idea of republic declaring it to be an un-Islamic concept, fearful of Ataturk's Turkey. Consequently in 1925 Reza Khan became Reza Shah, approved by the ulama and the Majlis, thus giving birth to the Pahlavi dynasty. Pahlavi was the name of the ancient Iranian language (chapter 2), which Reza Shah adopted as his surname, since the commoner Reza Khan had no surname of his own. Other Iranians, with existing surname Pahlavi, were forbidden to use it.

The objective of Reza Shah was to centralise the administration, modernise the bureaucracy and to strengthen the army. He had no interest in social reform and had no concern for the poor. All opposition was ruthlessly suppressed. Initially pliant to the clergy, he soon became strong enough to move against them, and proceeded speedily with aggressive secularisation, surpassing even Ataturk. The opponents were simply eliminated, starting with Ayatullah Mudarris who was his former staunch supporter. Mudarris was first imprisoned in 1927 and then murdered in 1937. The Shah's rule was brutal, but he centralised the administration and reformed the judiciary with a secular law (replacing Sharia law) for civil, criminal and commercial codes. He also attempted to modernise the country with new amenities (such as electricity) and to industrialise the economy, but government control stifled the economy and its growth. Wages were low, exploitation rife and innovation miniscule.

The reform programme of Reza Shah was superficial, as it tried to impose modern institutions on top of an old agrarian economic structure, a reform that failed in Muhammad Ali's Egypt. It was bound to fail in Iran as well, as it eventually did. There was no fundamental reform of the society. Agriculture in which 90 percent of the populations were involved was ignored, the poor were neglected; 50 percent of the national budget went to the army, and only 4 percent to education. As in Egypt, two nations developed, a small Western educated upper class, and the rest of Iran as the lower class, consisting of the uneducated backward masses dependent on the disgruntled ulama for guidance.

Islamic celebrations were replaced by Iranian national ones, taken from earlier history, such as the *nau-rouz* (New year) festival. Pilgrimage to Mecca (hajj) was forbidden and attempts were made to stop the mourning of the Ashura. Western dress was made obligatory for men, and the veil was banned for women. Veils were torn and cut into pieces with bayonets from the faces of veiled women by soldiers. Reza Shah ordered all government employees, from the road sweepers to the ministers, to parade their wives in public, while women were still denied the right to vote[1]. As men could not prostrate themselves during prayers with Western-style hats, they dem-

---

[1] Women did not have the right to vote in Europe either. Full voting rights of women (equal to that of the men) were granted in the UK only in 1928, following the efforts of the Suffragette movement.

onstrated against it. In 1935 in one such demonstration, many unarmed demonstrators were killed and hundreds wounded by soldiers in the holy shrine of the 8th Imam in the Mashad mosque. The Iranians came to dread secularism. Like Ataturk, Reza Shah wanted Iran to "look modern" and "civilised". But unlike Ataturk, he did not create democratic political parties as opposition; instead he imposed a tyranny.

Despite the special concessions he granted to the British on the Anglo-Persian Oil Company, Britain forced him to abdicate in 1941 after the start of the Second World War, because of his neutrality in the War. He was replaced by his 20-year old son Muhammad Reza, a weak but arrogant and extravagant character. For a while the ulama saw a chance to re-assert themselves, but not for long. In 1953, Shah was overthrown for his refusal to handover power to Musaddiq, the popular prime minister, who nationalised the Anglo-Persian Oil Company in 1951. At this point the American CIA got involved and engineered the downfall of Musaddiq and return of the Shah in a counter coup. Many opponents were hanged, but Musaddiq (who was old, infirm and related to the Qajars) was only jailed as he was politically too powerful. Most of the oil facilities were given back to a Western cartel. The Iranians felt betrayed and humiliated by the West, especially by the Americans. Subsequently oil revenues soared, particularly after 1973. The Shah spent much of it in grandiose projects and in creating family wealth. Corruption became endemic and misrule became the norm. His secret service agents terrorised intellectuals and dissenters, many of whom were physically liquidated. In 1979, the Iranians revolted, the Shah lost his throne and left Iran for ever. Ayatullah Khomeini came to power. Modernisation met its death long before, ironically at the hands of the moderniser Reza Khan.

12.7 Kemal Ataturk of Turkey (1881-1938)

Much of Greek mythology was entwined with the coast of Anatolia and equally much of ancient Greek science originated in its coastal islands, which perhaps explain why Turkey (Anatolia) prides itself as being a European country, even though geographically most of it is in Asia (except for its capital Istanbul and its immediate surroundings). Mustafa Kemal, who originally drew his inspiration from Islam, later concluded Islam to be the main stumbling block to modernisation. Kemal, the nationalist, force-fed the sick man of Europe a heavy diet of European modernisation, but after some 80 years, indigestion from over-eating and heartburn continue even to this day. No discussion on an Islamic revival can be complete without an examination of Kemalism.

Mustafa Kemal, born in Salonika in 1881 CE, was a member and later leader of the Young Turks military movement (chapter 11), actively engaged in the modernisation of Turkey. But it was the Young Turks who sided with Germany during the First World War, at the end of which Turkey lost all its colonial territories. Nevertheless, Mustafa Kemal emerged as the supreme military leader, who managed to beat the Greeks in 1919 and saved what is now modern Turkey from subjugation. He deposed the sultan, abolished the Khelafa and became the President of the new democratic republic of Turkey.

Kemal saw the salvation of Turkey in rapid Westernisation, which he believed would be hampered by Islam. He therefore identified Islam as the biggest cause of backwardness, which he removed declaring Turkey a secular state, independent of Islam, and he replaced Sharia law and Sharia courts with their European equivalents. Swiss family law was adopted, women were given equal rights and madrasas (religious schools) were closed. Furthermore, the Arabic script of the Turkish language was replaced with Roman script in 1928 and surnames were made compulsory in 1935. He banned *Adhan* (the melodious call for Muslim prayers) in Arabic, and declared the wearing of the Fez (cap) as a criminal offence, but approved bowler hats as part of "civilised" Europeans attire, an approval that might amuse some Europeans today. For good measure, he legalised the sale of alcohol and ball-room dancing (both his favourites). After achieving all these changes he proudly announced in the Grand National Assembly in 1932: "We are by now a Western nation".

To signal a complete break with Islam, he gave up his Muslim sounding name Mustafa, and took up the name Kemal *Ataturk* (Father of the Turks). However, the transformation from theocracy into a modern secular state had a few hiccups. He was able to crush religious opposition, though not always by democratic means. In the constitution he imposed, he had given the army the task of ensuring the secular nature of the Turkish state, which has led to many military coups in recent times.

When he died in 1938, he was buried as a Muslim on the insistence of his sister. In 1954 she held prayers in the famous Sulaimanya mosque in Istanbul for the peace of his soul and the absolution of his sins [Zak89/p210]. People sometimes come to religion in a mysterious way. According to some recent reports, even Stalin apparently confessed to a Russian Orthodox priest in great secrecy at the height of the Second World War, when the Soviet Union was at the greatest danger of collapse.

Kemal Ataturk thought he had created a secular modern European state, but no sooner had he been buried than Islam raised its head again. In the 1970's some of the religious bans were lifted. Since then, religious education is being permitted even at university level, Imams are being trained, mosques are being built, *Adhan* has been reverted back to Arabic and the ban on the hajj has been removed, but the state has remained secular. Recently the army threw out a duly elected Islamic Party from Government, but now a new Islamic Party is in power, which however recognises the secular character of the state.

But has Turkey achieved modernisation? Has it developed the rule of law, human rights, and science & technology? Because of its human rights record, Turkey is finding it hard to become a member of the European Union. As regards science, the record of Turkey is not significantly better than other Muslim countries.

It is tempting at this point to think about the former Soviet Union. After seventy years of anti-religious communism, religions have been revived in all its former republics with an astonishing speed. Turkey is no different. Modernisation cannot be achieved by disbanding religions via decrees. It instead has to proceed through enlightened evolution, as has happened in Western Europe. The people should always have a choice, and be given freedom to choose. The reformers need to win both hearts and minds, and must demonstrate convincingly how the reform would work and bring practical benefits.

Conclusion

Two lessons can be learned from the reformist rulers. Modernisation cannot be forced from the top without some fundamental changes in the nature of the society, and a God-less creed would not succeed in

Muslim countries. The reform has to be channelled through religious reformation, for which we need our reformist scholars.

Such a reformation could be provided through what Pervez Hoodbhoy [Hoo91] labelled as "the reconstructionist approach", in which anti-science and anti-modernism of the orthodoxy is replaced by a re-interpretation of the faith in order to reconcile the teachings of Islam with modern science. This school of thought holds that Islam practiced by the Prophet of Islam and the Rashidun was revolutionary, progressive, liberal, and rational, and that Islam subsequently slid into stultifying rigidity and reactionary dogmatism through the practice of *taqlid* over *ijtihad*. The reinterpretation of scripture involves the search for greater truth as advocated by the Mutazilites and also by some later and recent reformers. Both Sir Sayyed Ahmad Khan and Syed Ameer Ali belonged to this select group of reformers. The views of the more recent reformers are discussed in the next chapter.

Finally, Muhammad Ali wanted rapid modernisation, but unlike Ataturk he did not see the aping of Western customs and manner as the sign of a civilised existence. Kemal's superficiality influenced both Amanullah Khan and Reza Shah, as they all worshipped the Western exterior, without paying any attention to the deeper Western concept of human dignity, rule of law, individual rights, freedom of expression and representative governance, which are the essential prerequisites of modernisation. Any worthwhile socio-religious reformation has to incorporate such essential values if its practitioners are to build a science-based modern society. But for modernisation to proceed successfully in a harmonious environment, the reformation must also help change the Muslim self-image from the *Closed* to the *Open World Assumption* on knowledge. These issues are discussed in chapter 14.

References and Sources

[Ali67] Syed Ameer Ali: The Spirit of Islam, first published in 1922, reprinted by Methuen & Co. in 1967.

[Ali68] Syed Ameer Ali: The Legacy of Islam, first published in 1931, reprinted by Oxford University Press in 1968.

[Arm00] Karen Armstrong: The Battle for God, HarperCollins, 2000, p112.

[Bri71] Encyclopaedia of Islam, ed E. J. Brill, (Leiden 1971), Vol 3, p912.

[Ham62] Hamilton Gibb: Studies on the Civilisation of Islam, Princeton University Press, 1962, p142/3.

[Hal68] P. M. Halt (ed): Political and Social Change in Modern Egypt ... ... , London 1968.

[Hoo91] Pervez Hoodbhoy: Science and Islam, Zed Books, London, 1991.

[Hou83] Albert Hourani: Arabic Thought in the Liberal Age (1798-1939), Cambridge University Press, 1983.

[Itz76] Norman Itzkovitz: The Ottoman Empire, Islam in the Arab World, ed. B. Lewis, published by Alfred A. Knopf, New York, 1976, pp 273-300.

[Ked72] N. Keddie (Ed): "Scholars, Saints and Sufis: Muslim Religious Institutions in the Middle East since 1500", London 1972 [Many articles].

[Ked83] Keddie: An Islamic response to Imperialism, University of California Press, 1983.

[Ken75] E. S. Kennedy: Cambridge History of Iran, Vol 4, 1975, see p 591 for Safavid Sciences.

[Rah79] Fazlur Rahman: Islam, University of Chicago Press, 2nd ed. 1979.

[Rut91] Malise Ruthven: Islam in the World, Penguin 1991.

[Tro78] C. W. Troll, Sayyid Ahmad Khan, Oxford University Press, 1978.

[Vat91] P.J. Vatikiotis: History of Egypt, Weiderfeld and Nicholson, 4th ed 1991, pp255-256.

[Wat88] W. M. Watt: Islamic Fundamentalism and Modernity, Routledge, London 1988.

[Zak89] R. Zakaria, The Struggle Within Islam, Penguin, 1989.

CHAPTER 13
MODERN RETHINKING ON REFORM

We have presented in chapter 4 the development of the traditional views on hadiths and Sharia, largely formed between the 8th and the 10th centuries and has remained the basis of Muslim lives ever since. However as Muslims fortunes started to decline, particularly rapidly from the 18th century onwards, the traumatised Muslims began to search for an explanation for their plight. For the orthodox the reason for the decline was the lax observance of holy Sharia, while for others the decline was prompted by Sharia itself. In this chapter we shall revisit some of the issues and examine the ideas of modern liberal reformers, who saw the failure to create a science-based technological society, due to the practices of Sharia, as the major reason for the marginalisation of Muslims.

Following the irreversible decline of Muslim power throughout the world in the 18th century, a group of reformers came to the view that the answer lay in the rejuvenation of Islam through a reassessment and reinterpretation of sunna and hadiths, using *ijtihad*. Luminaries include Shah Wali Allah of Delhi (1702-62) who was affected by the breakdown (not yet collapse) of Mughal authority, and also Muhammad al-Shawkani (1760-1834) of Yemen. This purification approach, partly linked with the Saudi Wahhabi movement, gave rise to a new *ahl al-hadith* group of reformers who believed in a single literal meaning of the texts of the Quran and hadiths, which in their view provided salvation for Muslims. It is perhaps worth noting here that the rest of the Muslim world always viewed even the hanbalis, the relatively moderate parent madhab of the Wahhabis, as extreme (chapter 4); then, as now.

The 19th century spelled disasters for Muslims throughout the world. It witnessed the collapse of Mughal power in the face of the British colonisation of India, Western domination of the Qajar authority in Iran, the dwindling of the Ottoman might in Europe and the transfer of Egypt to British control. This trauma produced a more drastic reform movement in the early 20th century, both in India and in Egypt. These new reformers did not consider the *ahl al-hadith* doctrine to be radical enough to meet the new challenges of science, technology and Western dominance. They produced a new doctrine of *ahl al-Quran*, in which the Quran alone was perceived to be self-sufficient for Muslims, to the exclusion of hadiths, which they argued were unreliable anyway.

Both Sir Sayyed Ahmed Khan (1817-98) of India and Muhammad Abduh (1849-1905) of Egypt were concerned about the authenticity and reliability of hadiths, but they did not formulate the *ahl-Quran* doctrine which was postulated after their time. This doctrine was established in the early 20th century by the later reformers, such as Abd Allah Chakralawi (d 1930) and Khwaja Ahmad Din Amritsari (1861-1936), both from India.

It should be clear that both the *ahl al-hadith* and *ahl al-Quran* reformers were unhappy with hadiths, but while the *ahl al-hadith* supporters sought solutions through the re-study and re-interpretation of hadiths, the *ahl al-Quran* group rejected hadiths altogether. This rejection was based on two theological points: (i) hadiths are unreliable and unimportant and (ii) the Quran alone is complete and sufficient. We shall review these arguments below, along with some alternative points of view, revisiting in the process for the sake of completeness, some of the hadith issues already discussed in chapter 4. We begin with a review of the early Islamic practices leading to the development of Sharia and its uncritical support by the ulama over the centuries.

13.1 Review of the Early Practices, Sharia and Ulama

The early Muslims in the first and second centuries of Islam did not always give precedence to the sunna of the Prophet over those of his prominent Companions, particularly the Rashidun. They did not even identify sunna with specific hadiths. In addition they did not draw any rigid distinction among the various sources of authority, such as the Quran and the sunna [Bro99/p9]. Within that informal environment, Islamic laws were drawn from the Quran, known practices of the Companions (of the Prophet), local practices and consensus (*ijma'*), but not from any explicit hadiths (which were not collected at that time anyway). The ancient school of Kufa that followed the Companion Abd Allah Ibn Masud[1] (d 32 AH/653 CE), and of Medina that followed Khalifa Umar, operated on this basis, generally accepting a local practice as long as it did not contradict directly any Quranic provision. The pro-hadith group (i.e. *as'hab al-hadith* – the partisans of the tradition, as mentioned in chapter 4) came into existence in the second century of Islam, in around 770 CE (some 120 years after the death of the Prophet), demanding all laws to be based on explicit hadiths. The schools of Kufa and Medina initially resisted the pro-hadith approach of the partisans, but had to give in later, even though not completely, when at the end of the second century Imam Shafi'i (767-820) elevated the pro-hadith doctrine onto a level where hadiths were the only valid authority for interpreting the Quran.

The partisans equated the Quran with hadiths, stating that, "whenever Muhammad received a revelation, he was also delivered a sunna to explain it". Also "Gabriel used to descend to the Prophet with sunna just as he descended with the Quran". Sunnas were viewed by the partisans as *un-recited* revelation (*wahy*). Imam Shafi'i wrote: "Prophet of God proffered nothing that was not *wahy*" [Bro99/p16]. It was further claimed that "The sunna rules on the Quran, but the Quran does not rule on the sunna" and that "the Quran has greater need of sunna than the sunna of the Quran" – that is, in case of Quranic conflicts with the sunna, the sunna would prevail!

There is a hadith that seems to support the claim of Imam Shafi'i: "Whoever interprets the Quran using his own intellect, even if the interpretation is correct, he is committing a grievous sin" [Tirmidhi and Abu Daud]. Given the injunctions of the Prophet and the Rashidun on not writing down hadiths (see below) and given that in Islam intellect is believed to be the greatest gift of God to man-

---

[1] As the treasurer of Kufa, Ibn Masud opposed the misuse of funds by governor Walid, which to his surprise was not objected to by Khalifa Uthman, half-brother of Walid. Following this and several other disagreements with Uthman, Ibn Masud became disenchanted with Uthman and before his death, he left word forbidding Uthman to conduct his funeral service. He died in 653 CE, before Khalifa Uthman [Abb85/p109-11].

kind, how can one trust this hadith? Many experts believe that this hadith was manufactured to chastise the Mutazilites.

The pro-hadith ideas were opposed strongly in Shafi'i's own life time by the speculative theologians, known as the *ahl al-kalam* followers, who rejected hadiths altogether in favour of the Quran alone (mainly because of the unreliability of hadiths), and therefore opposed Shafi'i's *as'hab al-hadith* doctrine. Their demand, that a hadith can be accepted for the interpretation of the Quran only if its authenticity matched that of the Quran, disqualified all hadiths except for a very few. Their cause was later taken up by the Mutazilites who insisted on sourcing their religious system on the Quran alone, and used the Quran to discredit hadiths, even though they did not produce any system of law [sch64/p58-81].

Over the ensuing centuries the ulama were the carriers of the *as'hab al-hadith* approach – they were the promoters and upholders of Sharia, not its reformers. For them the defence of hadiths was part of a battle to preserve their own positions as the guardians and interpreters of hadiths, and hence the supreme authority of the religion, since hadiths are projected by them as the only valid (divine) interpretation of the religion. They banned any idea that could weaken their exclusive right to interpret and denounced anyone who could possibly threaten their way of interpreting the scripture, as they did with Taha Hussain (1889-1973) of Egypt [Hou83], who was blind from childhood but nevertheless later rose to become Minister of Education (1950-52) in Egypt. In 1926 he wrote a book on ancient (pre-Islamic) Arabic poems, in which he applied a text analysis technique that showed some of these poems not to be pre-Islamic. The reason why the ulama got so upset was that if such a technique were to be applied to the scriptural texts, then it might potentially cast doubt on their authenticity – and that is something the ulama could not possibly tolerate. The book had to be withdrawn. While he lost his job in one university, he got another job in the American University of Cairo, where his colleagues supported his freedom of speech[2], and where he later became the first Egyptian Dean of Arts.

---

[2] Recall when Raziq (section 9.2) was declared a heretic, he lost his livelihood and had to leave Egypt for Paris. In contrast Taha Hussain stayed in Egypt. Observe the cultural difference in three Egyptian universities: (i) a religious university which dismissed Raziq after declaring him a heretic, (ii) an Egyptian-managed Western university which dismissed Taha Hussain because the ulama declared him a heretic and (iii) the American University of Cairo which supported Taha Hus-

More recently the Egyptian writer Nasser Abu Zaid, Professor of Arabic language at Cairo University, argued that the Quran should be interpreted in the light of the 7th century Arab politics. The ulama were outraged, he was declared an apostate and a case was lodged against him in 1993 to separate him from his wife, since an apostate is a non-Muslim who cannot remain married to a Muslim woman. When the case was still pending, he left Egypt with his wife in 1995 for fear of his life. Hassan Hanafi, another Professor at Cairo University, was also charged with apostasy in 1997, even though he wanted to see Islam as a liberal religion with social justice, with God everywhere (hence the charge of pantheism against him). There is also the Egyptian feminist writer Dr Nawal Al-Saadawi, who was declared heretic in 2001 and her husband was advised to divorce her [Web04]. There are many other examples.

As we also have seen in the previous chapters, the ulama opposed everything that could dilute their hold on religion and supported anything that could enhance their hold – all in the name of Sharia. As they were uncomfortable with technology, they forbade the teaching of science and technology (including mathematics), they destroyed the new observatory in Turkey, they banned the introduction of printing, they prevented the establishment of a democracy in Iran, and they even executed a pro-technology sultan[3] in Turkey. They opposed the liberal policy of Akbar that led to the creation of the great Mughal Empire in India and supported the illiberal policy of Aurangzeb that led to the destruction of the same empire. They never hesitated in giving fatwas for killing the innocent Ottoman princes and even a sultan in favour of his son[4]. Their control over the Ottomans was so total that even the examination of a wrecked foreign warship required their permission in the form of a fatwa. They stifled reform and modernisation. For a thousand years, while the *as'hab al-hadith* approach dominated Muslim societies everywhere, Muslim fortunes crumbled everywhere.

---

sain's freedom of speech. The first case can be understood in terms of religious orthodoxy, but the second case is more disturbing – a denial of freedom of speech even in a secular university due to Sharia conditioning.
[3] Sultan Selim III, see chapter 11.
[4] Sultan Ibrahim, see also chapter 11.

13.2 Unreliability of Hadiths

It would appear that the Prophet disapproved of the writing down of his hadiths, presumably to avoid confusion with the superiority of the Quran. According to Abu Said al-Khudri: "The Prophet said: 'Do not write down anything from me except the Quran. Whoever writes down anything other than the Quran, must destroy it'. " [Muslim, and Hanbali].

Further support to this prohibition is provided by both Zaid bin Thabit (Prophet's secretary and the Compiler of the Quran under Abu Bakr, Umar and Uthman) and the hadith expert Ibn Shihab al-Zuhri (d 741). When Khalifa Muawiya asked Zaid bin Thabit to write down a story about the Prophet, he refused saying: "The messenger of God ordered us never to write down anything about his hadiths". The first person to write down hadiths (under order from Khalifa Hisham) was al-Zuhri who was reluctant to do so in view of prohibition in the writing down of hadiths [Bro99/p92]. This was some 100 years after the death of the Prophet. Furthermore, Khalifa Abu Bakr forbade the collection of hadiths, when he found the Companions issuing contradictory hadiths. Khalifa Umar chastised some Companions (in particular Abu Huraira) for spreading hadiths. Khalifa Ali also disapproved the writing of hadiths. Muawiya was the first Khalifa who started search for hadiths, as mentioned in chapter 4.

A cardinal basis for the belief in transmitted hadiths is that the Companions of the Prophet were trustworthy, that they did not lie regarding hadiths and that they correctly stated what they have actually heard from the Prophet. But it seems that many Companions were not so trustworthy. There is some evidence that the Prophet himself did not trust some of his Companions concerning hadiths, as he is reported to have said: "Let whoever deliberately tells lies about me, take his place in hell".

Furthermore, some Companions have become sources for larger numbers of hadiths than others. Abu Huraira who was converted to Islam in the Hijra 7 and spent under three years with the Prophet, is the biggest source of hadiths (5376 hadiths)[5], much of which he pro-

---

[5] As stated in chapter 4, al-Bukhari has hadiths from 208 and al-Muslim has from 213 Companions, of which 149 Companions are common. However, each has accepted only a small number of hadiths from each Companion as genuine. For instance al-Bukhari accepted only about 142 hadiths (out of possible 2210) from Ayesha. The rejection does not necessarily imply that the Companions were lying,

duced under Muawiya, having previously been chastised and forbidden by Umar. This was the highest number of hadiths anyone had produced, and many of these hadiths were contradictory. This prompted even Ayesha (the wife of the Prophet) to call him a liar. He was also criticised by Abdullah Ibn Umar and Ibn Abbas for his numerous hadiths. The next closest Companion in terms of hadiths was Abdullah Ibn Umar, who had 2630 hadiths. While Ibn Umar wrote down his hadiths, Abu Huraira did not, boasting of an infallible memory. Ayesha (2210 hadiths) attacked Anas (2286 hadiths) for transmitting hadiths though he was only a child at the time of the Prophet's death. Hasan (grandson of the Prophet) called both Ibn Umar and Ibn Zubair liars. In contrast Abu Bakr was the source of only 132, Umar 537, Uthman 146 and Ali 536 hadiths [Sid61]. Fatima died soon after the Prophet before hadiths became important. However, it should be noted that most hadiths had been rejected by both Bukhari and Muslim, as being unreliable, possibly forged by some later transmitters using the names of the famous Companions.

Bukhari apparently claimed to have known 600,000 hadiths [Ahm97] and accepted only about 4000. If it took Bukhari only 20 minutes to collect, read, investigate, evaluate and record a hadith, then it would have taken him 200,000 hours, which is 66 years with 8 hours a day, every day of each year. In fact the investigation of a hadith would often take many days, since the sources were distributed over many cities, such as, Mecca, Medina, Damascus, separated by several days of camel-journey. Bukhari's total working life was said to be 40 years, implying that he could not have examined all 600,000 hadiths thoroughly even in terms of his own criterion which according to many scholars was not sufficiently rigorous (section 4.1). Therefore this claim of 600,000 if at all made cannot be true.

It is said that Imam Ibn Hanbal claimed the existence of seven million authentic hadiths[6] [Ahm97], which if true, would have taken the Prophet to utter one hadith every 78 sec, 18 hours a day (if he did nothing else) over his prophetic life of 23 years! Again if the Imam had actually made this claim, most of these hadiths would have been forgeries, even if he did not know it. We can perhaps safely conclude

---

it could be that some later transmitters had fabricated hadiths using the names of the Companions.

[6] However only 7000 hadiths from some 400 Companions had been included in the Hanbali collection.

that these numbers of 600,000 and 7 millions are highly exaggerated by some later enthusiasts, but the question remains: how far can one trust the reliability of the collected hadiths? (see also chapter 4).

There are many contradictory hadiths from the close Companions of the Prophet, all of which cannot be true. It is not that they were deliberately fabricating hadiths, mostly it was perhaps a question of who heard what and when and under what context, how they have interpreted what they have heard and what they remembered. If one asked a set of witnesses to describe an incident, or a conversation, that took place a few years previously, the accounts would vary. It is worth stressing that at the time of the Prophet, his hadiths were not given the same status as that of the Quran, in the sense that they were not carefully preserved – they remained as oral memory in the minds of the hearers, without any attempt to verify their truth and accuracy from several contemporary hearers, as was done with the Quran.

If the Prophet had said something important, then he must have said it to multiple Companions (possibly during the Friday Khutba). In particular those close to him, such as, Abu Bakr, Umar and Ali would be expected to know about it. On this basis, one would expect a genuine hadith to be sourced at least by multiple Companions, even if they are not the close Companions. In fact a genuine important hadith should have not only multiple-Companion *isnads*, but also these *isnads* should be independent of each other – that is, these *isnads* should not have any narrator in common. However we do not know how many hadiths have such independent multiple-Companion *isnads* (IMCIs)[7]. Unfortunately IMCIs can also be manufactured by forgery and hence would not on their own guarantee genuineness.

Most of Abu Huraira's hadiths have single *isnads* and are often mutually conflicting. It may be that some of these hadiths are not really by Abu Huraira, but others have made them up in his name later. How can anyone know whether a Companion is truthful or whether a subsequent narrator had made up a false hadith using the Companion's name as the source. Checks on piety and memory

---

[7] The Bukhari and the Muslim collections, for example, have hadiths (i.e. *matns*) with multiple Companions as the sources, and therefore it should be possible to find out how many *matns* have IMCIs. However, there are other issues as well, for example, some of these *matns* (in Bukhari and Muslim) conflict with well established sunnas and some do not even make sense [Kam04, pp 201-220].

(chapter 4) do not provide any guarantee of correctness. It seems hadiths have been accepted on good faith rather than on any rigorous rational criterion.

Some Muslim scholars see a parallel between the writing of the Bible and that of hadiths. The life of Jesus in the New Testament was based on narratives produced by some early Christians long after the death of Jesus. Fundamentalist Christians believed these narratives to be divinely-inspired and hence literally true (despite many contradictions among these narratives), while other Christian theologians are sceptical. Likewise, hadiths were written and collected long after the death of the Prophet regarding what he had said and did. Similarly orthodox Muslims believe the collected hadiths to be true (despite some contradictions), while other Muslims are dubious. A similar parallel also applies to the Old Testament.

Muslim scholars agree that hadith forgery took place on a massive scale [Bro99/p93]. Was it good enough to use only the *isnad* verification to differentiate between the valid and forged hadiths, without checking the content? Furthermore, the *isnad* verification technique itself was also questionable. Only a small number of hadiths at best can be traced with any degree of certainty to the Prophet [Bro99/p104]. Imam Abu Hanifa approached hadiths with the assumption that very few of them can be proved to be *sahih* – an assessment to which both Malik and even Shafi'i apparently agreed. Even the conservative Pakistani ulama Abul Ala Maududi was concerned about the unreliability of hadiths and advocated the re-examination of their authenticity by modern means.

13.3 New Reformist Thinking

We have discussed in the last section (more fully than in chapter 4) the reasons why hadiths are considered to be unreliable by both *ahl al-hadith* and *ahl al-Quran* reformers. However, the *ahl al-Quran* reformers regard hadiths to be unimportant anyway and the Quran to be the exclusive document for the guidance of Muslims, as explained in this section. There is also a post-*ahl al-Quran* view, in which some messages (ayahs) of the Quran are considered to be more eternal than others, as advocated by scholars such as An-Naim of the Sudan and covered later in this section.

The *ahl al-Quran* followers used hadiths to prove that hadiths are unnecessary and unreliable, and also used verses of the Quran to prove that hadiths have no status as religious texts. As we have seen in the previous section, there are hadiths stating that the Prophet himself prohibited the writing down of hadiths. In addition, not only did the revered three Khalifas ban the writing of hadiths, but many other well-respected Companions also refused to write down hadiths because of the Prophetic injunction. To the *ahl al-Quran* supporters, this ban implies that hadiths were not important enough to be preserved accurately and were not an essential part of the religion, even if genuine hadiths can be found. According to them, there are further reasons against the use of hadiths, as discussed below.

It is reported that Khalifa Umar on his deathbed instructed the Muslims to follow (after he was gone): the Quran, the *Muhajiruns* (the Muslims who emigrated with the Prophet from Mecca to Medina) and the *Ansars* (the Medinans who welcomed the *Muhajiruns*), the people of the desert and finally *ahl al-dhimma* (the protected communities of the Jews and Christians). Neither the sunna of the Prophet nor hadiths were mentioned, which provided a basis for the *ahl al-Quran* claim that sunna and hadiths were not important. This point is further reinforced by the fact that the early Islamic legal reasoning appears to be virtually hadith-free [Bro99/p11]. The first known reference to sunna and hadith was made when Khalifa Abd al-Malik (65-86 AH) asked Hasan al-Basri to give a legal opinion on the doctrine of the Qadarites (chapter 3); but even there, sunna was mentioned only in a vague manner; no hadith was cited at all. The *ahl al-Quran* supporters argue that these facts constitute further and sufficient proof that hadiths were not considered important – not important enough even to be written down. Since sunna was not as carefully recorded as the Quran, surely it implies that it was redundant, they contend.

They also claim that the term sunna of the Prophet is post-Quranic (since the Quran does not speak of the sunna of the Prophet), even though it (the Quran) does describe the Prophet's character as being exemplary and great, and therefore, by implication, to be followed. This clearly does not imply that hadiths provide the only possible interpretation, let alone the divine interpretation, of the Quran – even where the hadiths are reliable. The ancient legal schools did indeed follow the Prophet's example, but still were opposed to the *as'hab al-hadith* group. Even Shafi'i's own work showed that in the minds of the early Muslims, the sunna of the Prophet was only one of several sources of religious authority, such as the sunna of the Companions and the sunna of the Rashidun. There are even bigger problems with hadiths and their authenticity, they assert. They claim that by the time hadiths came to be compiled into written volumes (such as by Bukhari) in the third century of AH, they were already damaged beyond any reasonable hope of restoration.

They also argue, that each hadith transmitter, even when totally honest, transmitted only what he understood, not necessarily the actual words of the Prophet. Thus a series of persons successively producing what they understood, could not be taken as the actual words of the Prophet. While the Quran was carefully protected, there was no attempt by the early Khalifas to protect the accuracy of hadiths. Gathering of hadiths started from about 100 years after the death of the Prophet, when fake hadiths were already in abundant supply. Furthermore, even the collection of Bukhari, the most respected collection, checked only *isnads,* not the contents, and even then only on the basis of the character of the transmitter, which was in turn based on the wholly unreliable assumption that a pious person could not produce any corrupt hadiths. They claim that even the best collections, such as those of Bukhari and Muslim, are mixtures of truth and falsehood, riddled with blatant blasphemies and absurdities. For instance in Bukhari there are hadiths that are (i) absurd, such as Moses punching the eyes of *Ajraeel* (the angel of death), (ii) repugnant, such as sexual details of the Prophet, and (iii) vulgar, such as the devil fleeing farting. How could such a collection be reliable, authentic and respectable? – they argue.

There are ayahs in the Quran which can be used to dismiss Prophetic hadiths, for example:

"These are God's revelations that we recite for you with the truth. In which hadith besides the revelations of God do they believe?" Quran [45:6].

This ayah implies that the only valid hadiths in God's eye are His messages in the Quran. There are also other ayahs that relate directly to decision-making procedure, for example:

"... Therefore, congratulate My servants who listen to all views, and then follow the best. These are the ones guided by God – these are the intelligent ones."
Quran [39:17-18]

implying that decisions should be made in consultation. Nowhere in the Quran does it ask for decisions to be made on the basis of sunna or Prophetic hadiths [Ahm97].

The *ahl al-Quran* group was strengthened later by the contributions of Muhammad Aslam Jayrajpuri (1881-1955), a former *ahl al-hadith* scholar. Similar ideas were also published in 1906 by Muhammad Tawfiq Sidqi (an associate of Rashid Rida (1865-1935)) of Egypt, who was later forced to recant for fear of upsetting the all-powerful ulama. In the 20th century there were new writers, such as Ghulam Ahmed Pervez in Pakistan and Mahmud Abu Rayya in Egypt, who have developed sophisticated arguments in favour of the *ahl al-Quran* doctrine and against the acceptance of hadiths. Their central argument is that the Quran is self-sufficient and complete, it needs no external aids (except the knowledge of Arabic) to interpret it, as stated in the Quran itself "the book that explains all things" [16:89]. Also God Himself bears witness that He has "omitted nothing from the Book" [6:38]. Therefore, they argued, what is obligatory for man does not go beyond the holy Book. Hence no sunna nor any hadiths are needed.

Parvez, who was a contributor to Ayub Khan's Constitution of Pakistan, cites several verses from the Quran which corrected the decisions made by the Prophet:

1. Against the decision of the Prophet about prisoners at the battle of Badr: "It is not fitting for the Prophet that he should have taken prisoners of war until he has thoroughly subdued the land". Quran [8:67].
2. After the Prophet approved a request for exemption from military duty: "God forgave you! Why did you exempt them before those who were truthful, ... and you knew the liars". Quran [9:43].
3. When the Prophet vowed not to eat a particular food: "Oh Prophet, why do you call forbidden what God has declared permissible". Quran [66:1].

These verses and their contexts imply that the Prophet as a human being was fallible. Furthermore, the *ahl al-Quran* followers ask, if God has given him detailed guidance on every matter, why did He command him to consult his Companions? Based on these analyses they come to the following three decision-levels regarding the Quran and the sunna: (a) the Quran is divinely revealed and divinely protected against error, (b) the Prophet's decisions as the leader (amir), made in consultation with his Companions, were binding on his contemporaries as the central authority to execute Quranic commands, but ceased to be binding after his death since there will be new amirs who will make their own decisions, and (c) the Prophet's personal decisions were subject to errors and were never binding even on his contemporaries.

This is indeed a radical departure and outlook. They claim that this approach represents the true intention of the Quran and the Prophet, and had been implemented during the time of the Rashidun who did not follow the Prophet slavishly (see chapter 4). They blamed the ulama for misrepresenting this true Prophetic intention. Tawfiq Sidqi lends support to this view by arguing that sunna was meant only for the Arab people of that time as it was based on the local customs of the Arab people at that time. It was not meant to be binding on all Muslims for all time – if it were the Prophet would have asked it to be written down and preserved. Similar arguments have also been advanced by Nasser Abu Zaid (section 13.1).

At this point we shall digress a little to bring in the view of Fazlur Rahman[8], a well-known Muslim scholar. He admits that hadiths had been fabricated on a massive scale, but he does not reject hadiths in principle. For him genuine hadiths have an important role to play in drawing our attention to the sunna of the Prophet, sunna being crucial for Muslims. He probably captured the general mood of the reformers with his view that the decisions and precepts of the Prophet were generally ad hoc and informal in nature, and that he (the Prophet) applied them differently in different situations with discretions, sometimes after consultations with the senior Companions. Fazlur Rahman believes that Muslims should *not* study hadiths for direct application but only for clues to the spirit behind each hadith in a situational context. He opposes any formulistic or literalistic application of hadiths. According to him [Rah79/p251]:

---

[8] Fazlur Rahman was the Director of Central Institute for Islamic Research, Karachi, established by President Ayub Khan in the 1960's to provide a modernist interpretation of Islam for the regime, and for writing the Constitution for Pakistan. He later became a Distinguished Professor at the University of Chicago where he worked until his death in 1988. He was a renowned Islamic scholar – every sentence in his writing is precisely expressed and exudes his profound scholarship and command. We have also quoted him previously.

"What is necessary is to know the genesis and evolution of a given hadith in order to reveal what *function it did or was supposed to perform and whether Islamic needs do still demand such a function or not*. ... The thinking Muslim has to go right behind the early post-Prophetic formative period itself and to reconstruct it all over again. And this is exactly what the conservatives, who still largely control the mainsprings of power in the Community, not only refuse to do but completely fail even to recognise the need to do".

Clearly he does not believe hadiths to be the divine interpretation of the Quran, nor does he subscribe to the literal interpretation of hadiths. He is asking Muslims to seek the spirit behind a sunna and to apply it judiciously in situations where it is appropriate. This would imply that some hadiths may not be applicable, a point he made elsewhere while admiring Khalifa Umar's decisions (section 4.2) which went clearly against the sunna of the Prophet. Thus although he was not an *ahl al-Quran* follower, there is a substantial convergence of their views with his, particularly with respect to his advocacy of enlightened abstract (higher-level) interpretation and of judicious application of holy texts differently for different societies at different times. He was also unsparing on the inappropriate application of Quranic verses from narrow interpretations, such as to apply the Quranic ban on usury to bank interests [Rah79/p265]:

"It is sheer confusion to identify the system of *riba* (usury) banned by the Quran with that of modern banking and interest, which is a special device within the context of a modern concept of 'development economy' with an utterly different function".

He then quotes the Quran to oppose these ritualistic interpretations: "Vainglory can be no substitute for truth": Quran [53:28]. However, while Fazlur Rahman expresses the right sentiments for reform in terms of the interpretation and application of scriptural texts, the *ahl al-Quran* supporters provide a distinct doctrine that can be applied in order to propel reform forward.

There is another issue which lies at the core of the *ahl al-Quran* doctrine as regards the interpretation of the Quran. The *ahl al-Quran* followers say that whenever a divine word is revealed in a human created language for the understanding of a people, it becomes limited by that language and by the culture and constraints of that people at that time. Therefore the human interpreter of that divine word has to search for the eternal essence of that message, shorn of decorative elements, and then express it in a language at a given time for a given culture (see the next chapter for more details). Furthermore, according to them, the Quran provides a basic set of principles and is a general guide for moral behaviour, not a detailed prescription, which must be worked out according to circumstances. God never intended, they argue, for every detail of life to be eternally established by revelation. These views are closer to the Mutazilite doctrine of Khalq-i-Quran (chapter 3).

A more radical approach, advocated by Abdullahi Ahmed An-Naim of the Sudan [Ann96], goes even further, and can perhaps be described as a post-*ahl al-Quran* doctrine. An-Naim, who studied in Cambridge and Edinburgh, is a distinguished Professor of Law at the Syracuse University, with many laurels to his credit. His doctrine is based on the thoughts of Mahmud Taha, the founder of an "unorthodox reformist" movement in the Sudan, called the Republican Brotherhood. Taha was executed for his views in 1985, by President Numairi of the Sudan.

An-Naim makes a distinction between the Quranic verses revealed in Mecca as against those revealed in Medina. He claims that the Meccan verses are eternal and therefore provided an excellent basis of Islamic social law (public law), consistent with the requirements of universal human rights. Medinan verses, he argues, were meant to be for use in that society at that time and hence should not be applied in modern times. In [Ann96] he discusses the major issues in public law, including modern constitutions, criminal justice, international law, slavery (which Sharia discourages but does not actually forbid) and equal rights of women and of non-Muslims in a Muslim state – thus advocating a modern approach. Importantly he would *not* call his approach secular, but Islamic (new Sharia), based on the early Quranic verses. He does not mention hadiths, but then he is unlikely to accept a hadith that does not contain an eternal message.

There are arguments that one can offer in favour of this distinction between the eternal and non-eternal verses in the Quran. Referring to the Quranic verses [08:01] and [08:41] on war booty in section 4.2, one can argue that Khalifa Umar, in his decision to acquire all conquered lands for the state (instead of distributing the lands to the soldiers), had opted for the first verse as the eternal one, holding the higher truth. This apart, as we concluded in section 4.2, Khalifa Umar did not always follow the Quranic injunctions literally.

An important modern case for a distinction between the eternal and non-eternal verses can be found in forensic evidence for rape allegations. According to Sharia law, a rape accuser must produce four male (or eight female) witnesses to the actual act of penetration, otherwise the accuser will be punished for "false accusation". This law is based on a Quranic verse and hence does not acknowledge any DNA evidence, even if available. As a result, many rape victims are rotting in the jails of some Muslim countries where Sharia law is applied to rape trials. Should such witnesses be required today when a much better technique for determining guilt is available? The only conclusion any reasonable Muslim can arrive at is to treat this Quranic injunction as non-eternal and hence non-binding. Therefore there is a strong case for the separation of the eternal from the non-eternal verses in the Quran.

Controversy between the secularists and non-secularists, as to whether Islam is a private religion or whether it has also a sociopolitical role, continues even among the *ahl al-Quran* followers. There has not been any significant debate in Islam on secularism since the work of Ali Abd al-Raziq advocating the separation of state from religion (section 9.2). Taha Hussain of Egypt was a great reformer, he was partial to the doctrine of Khalq-i-Quran and he wrote:

> "Religion exists to give comfort to the hearts of men, by teaching certain general truths about the universe in powerful and moving symbols. These symbols must be judged by results, by whether they do in fact strengthen the individual or the nation. But men's mind change from age to age, the symbols must be expressed anew."

as quoted in [Hou83/p332-4]. According to him the distinguishing mark of the modern world is the virtual separation of religion and civilisation, which can be achieved by taking the "bases of civilisation" from Europe, without taking its religion. He urged the Egyptians to "follow the path of the Europeans so as to be their equals and partners in civilisation". His secularist identity is obvious.

13.4 Concluding Remarks

From the material presented in this chapter, we can identify four distinct contributions to modernisation of Islam, made by Fazlur Rahman, the *ahl al-Quran* group, An-Naim and Taha Hussain. There is a degree of convergence amongst them, which permits some synthesis at a certain level of abstraction. It is unlikely that the *ahl al-Quran* supporters would object to the view of Fazlur Rahman on how scriptural texts should be interpreted and applied, except that they are far clearer on seeking the eternal essence of a message, given the limitation of a human created language to capture the essence of divine message without elaboration. An-Naim is also concerned with that eternal essence of God's message for which he has proposed the separation of eternal and non-eternal messages in the Quran. Equally Taha Hussain speaks of general truths that give comfort to the hearts of men, but such truths can only be found by reinterpreting the symbols (i.e. the Quranic verses) from time to time, which is equivalent to finding eternal truth as understood by the human mind at different points in history. Thus we have a sound common foundation for reform.

As regards the need to reform, it cannot be expressed better than in the words of Fazlur Rahman [Rah84/p132]:

> "No structure of ideas can ever hope to make good or even command respect for a long time – let alone be fruitful – unless it is in constant interaction with living growing stream of positive and scientific thought. It is a sheer delusion to imagine that by stifling free positive thought, one can save religion from doing so, religion itself gets starved and impoverished."

It is unlikely that Taha Hussain, or any of the others mentioned above would disagree with this reformist statement. But reform will not be, and never is, easy. According to Daniel Brown, there is another dimension to all these talks of reform [Bro99/p59]:

> "At a deeper level, the controversy is also about the human interpreters of the Quran and their authority. If the sunna is the essential tool for understanding of revelation, then the experts in sunna are likewise indispensable. But if the ability to contextualise revelation is needed, then those who knew the modern

world will be the most able interpreters of the Quran and the knowledge about the tradition will be counted as superfluous. The deep sociological rifts between the traditional religious leadership and the western educated intelligentsia, between the religious scholars and the technocrats are thus projected onto the spectrum of modern Muslim attitudes towards the Quran and its interpretation."

This battle will probably continue for some time. But the point is that there are some Muslim theologians (*not* non-Muslim Western scholars), who are pious Muslims but at the same time see a different radical path within Islam itself, away from the backward-looking orthodoxy, as discussed in this chapter. Their approach seems to be the one that can create an environment in Muslim society for science and technology to flourish as mainstream Muslim activities. As long as the orthodox dominate the religion, science will remain outcast, at best tolerated as a fringe activity outside the comfort zone, in which good Muslims should not be engaged. So how should we proceed to build a science-based Muslim society? Time to proceed to our final chapter.

References and Sources

[Abb98] Nabia Abbott: Aishah – the Beloved of Muhammad, Saqi books, London, 1998.

[Ada76] C. C Adams: The Authority of Prophetic Hadith in the eyes of some Modern Muslims, edited by D. P. Little, Essays on Islamic Civilisation, E.J. Brill, Leiden, 1976.

[Ahm97] Kassem Ahmad: Hadith – A Revaluation, published by Universal Unity, ISBN 188 189 3022, Paperback, 1997. Also available in the Internet: www.free-minds.org. His five lectures on this topic in a Malaysian University was published as a book in 1987, later banned, and then translated into English with some improvements and published in 1997. This is worth reading.

[Ann96] A. A. An-Naim: Towards Islamic Reformation, Syracuse University Press, 1996 (ppb). This book is intended to provide the intellectual foundation for a total re-interpretation of the nature and meaning of Islamic public law, based on some ideas of Mahmud Taha (executed by Numairi in 1985) the founder of an "unorthodox reformist" movement in the Sudan, called the Republican Brotherhood. This book is not a dispassionate presentation but an advocacy for reform.

[Azm68] Mohammad Mustafa Azmi, Studies in Early Hadith Literature, Beirut, 1968.

[Bro99]: Daniel Brown: Rethinking Tradition in Modern Islamic Thought, Cambridge University Press, 1999. An excellent work, based on a PhD thesis, claimed to have been influenced by Fazlur Rahman.

[Bur77] John Burton: The Collection of the Quran, Cambridge University Press, 1977.

[Hou83] Albert Hourani: Arabic Thought in the Liberal Age (1798-1939), Cambridge University Press, 1983.

[Hou85] G. F. Hourani: Reason and Tradition in Islamic Ethics, Cambridge University Press, 1985.

[Juy83] G. H. A. Juynboll: Muslim Tradition, Cambridge University Press. 1983.

[Kam04] Muhammad H. Kamali: Hadith Studies, Islamic Foundation, UK, 2004.

[Rah79] Fazlur Rahman: Islam, University of Chicago Press, 2nd ed. 1979.

[Rah84] Fazlur Rahman: Islamic Methodology in History, Islamic Research Institute, Pakistan, [1965, 1984]. [This book might have a University of Chicago edition].

[Rut91] Malise Ruthven: Islam in the World, Penguin 1991.

[Sch64] Joseph Schacht: The origin of Muhammadan Jurisprudence, Oxford 1950, reprinted 1964, p58-81.

[Sid61] M. Z. Siddqi: Hadith Literature – its Origin, Development and Special Features, Calcutta University, 1961.

[Wat87] W. M. Watt: Islamic Political Thought, Edinburgh Paperback, Edinburgh University Press, 1987.

[Wat88] W. M. Watt: Islamic Fundamentalism and Modernity, Routledge, London 1988.

[Web04] www.cairotimes.com/content/issues/media/philos6.html, news.bbc.co.uk/2/hi/middle_east/1295075.stm.

CHAPTER 14

CHALLENGES FOR REVIVAL OF SCIENCE

We have presented in this book the sublime heights the Arabic science reached and the contribution it made to world civilisation. In the words of George Sarton the science historian, who has done so much to bring the achievements of Arabic science to the attention of the West:

> "The main task of mankind was accomplished by Muslims. The greatest philosopher, al-Farabi was a Muslim, the greatest mathematicians, Abu Kamil and Ibrahim Ibn Sinan, were Muslims, the greatest geographer and encyclopaedist, al-Mas'udi was a Muslim, the greatest historian, al-Tabari, was a Muslim."

The actual list is longer. The greatest physicians were Ibn Sina and Razi, and the greatest sociologist was Ibn Khaldun, besides other giants, such as al-Biruni, Ibn Rushd, Nasr al-Tusi and Ibn Shatir. It is unbelievable today that once the Muslims dominated the world both intellectually and materially, while today the world dominates the Muslims. The early Abbasid Muslims followed the Quranic injunction to seek knowledge and expanded its frontiers, while the later Muslims used selective hadiths, first to disapprove and eventually to extinguish it. Many great scientists had to sacrifice their aspirations, reputations, livelihoods and sometimes even their lives at the altar of orthodoxy – the history is bleak. The Quran says:

> "God would never withdraw the blessings He has bestowed on a people unless they change their inner selves". Quran [8:53].

A Muslim can surmise from this ayah that the blessing of the Golden Age was withdrawn when Muslims changed themselves into an anti-knowledge, anti-rational, anti-intellectual and anti-science bunch. The return to more orthodoxy exacerbated rather than alleviated the problem. If anyone has any doubt as to the reason for the failure, we again quote from Sarton who cared greatly for Arabic science [Sar50/p37]:

> "The decadence of Islam was not due to the lack or to the loss of material power and political supremacy but to the fact that the spiritual power within it has failed."

In this final chapter, we shall revisit some of the issues and discuss how Muslims can move forward in science from our current state of stagnation. For this we shall examine the general environment that affects the growth of science and then focus on the specific obstacles in Muslim society that need to be overcome, with ideas as to how we might be able to proceed with scientific revival.

14.1 General Environment

Drawing from chapter 9, we can identify the principle factors that facilitate the progress of scientific and technological developments, as listed below:

- Sympathetic Social Attitudes
- Infrastructure (Higher Education, Collegiality, Journals)
- Funding and Drive

All Muslim countries are developing countries, and most developing countries have sympathetic attitudes towards science. Exceptions are some Muslim countries where the religious orthodoxy discourages the study of science and indeed of any secular knowledge outside Islam. The last two factors bulleted above are relevant to all developing countries. Given the existence of extensive international collegiality and journals today, the presence of national collegiality and journals are not essential for the progress of science, but there is a need for high quality institutions of higher education, backed by adequate library facilities. The Internet, despite being a great help, is not a substitute for a good local library. It would appear that there is not a single internationally respected university of technology in any Muslim country, even though some Muslim countries are rich enough to support such universities.

Substantial funding and major initiatives also help rapid progress in science and technology, as has been demonstrated in the USA through the Manhattan Project (the Atom Bomb project) in the 1940's and later the space research initiative. It is perhaps worth mentioning that if one looks at the photograph of the world's top physicists[1] in 1911, there were none from the USA. It was customary for budding US scientists to visit Europe to complete their postdoctoral education, as did young James Watson (of DNA fame). One US postdoctoral researcher, in the late 1920's on his mandatory European trip, lamented the lack of any reverse traffic from Europe to the USA and he wished for a day when he might see at least some European physicists visiting the USA to do research. He was Robert Oppenheimer, later the father of the Atom Bomb and much later the Director of the Princeton Institute for Advanced Studies, where he controlled even Einstein's research plan. When the Manhattan Project was successfully completed, the Europeans were queuing up to work in US research centres. The traffic has reversed, it would seem, for ever. Rich Muslim countries can learn lessons, if they really wanted to support research. But then as Ziauddin Sardar, a well-known Muslim science writer in the UK, said [Sard77/p171]:

> "There are strong tendencies to withdraw to positions of nostalgic irrelevance, to cultivate dreams of bygone ages, and to refuse to think of the future and to be part of new solutions".

For Muslims, the absence of sympathetic social attitudes is obviously the most crucial handicap. This absence needs to be reversed

---

[1] A photograph of the top physicists attending the Savoy conference in Brussels hangs at the lobby of the Brussels Metropole Hotel, where the conference was held in 1911. The two most important physics theories of the 20th century, namely Relativity and Quantum Uncertainty, were European (in fact German), the debate on the validity of the Uncertainty Principle that was conducted in the late 1920's was entirely dominated by the Europeans – an exclusivity that is unlikely to be repeated in the 21st century. In fact, after the Manhattan Project all sciences, including physics, are largely dominated by the USA. The European physics research centre CERN (in Geneva) is an exception, which is run as an international centre, with many American scientists working in it.

by reforms as discussed in this chapter. Another major disincentive for any creative work in the Muslim world today is authoritarian institutions and religious orthodoxy, both of which curtail freedom of thought and action, and can only be resolved through political and religious reforms, as discussed below. Even assuming that Western educated Muslims will in general be favourable to reforms, religious reforms will not work without the participation of the ulama, who are the only people who can influence the masses and thus create essential socio-religious integration. This will require a reform of madrasa education.

However, to make this chapter self-contained, we shall start with a brief review of the rise and decline of Arabic science and conclude it with a discussion of how these reforms can provide the foundation for building a productive scientific environment.

14.2 Review of Rise and Decline

The Golden Age of Islam began with the Abbasid Khalifa Mansur and continued until Khalifa Wathiq, with a great boost of activity under Rashid and an even greater one under Mamun. The royal patrons and the scholars believed it to be their religious duty to seek knowledge to reveal the mystery of God's creation, following Quranic encouragement and hadiths. They gathered existing knowledge from all corners of the globe, translated the written sources into Arabic, and then set about improving upon that knowledge.

With religious duty providing their singular incentive, they excelled in everything they touched. They were aided by the fortunate coincidence of three factors: (i) religiously inspired scholars, (ii) a favourable political environment and (iii) tolerable religious environment. Strictly speaking (iii) was dependent on (ii), the toleration lasting as long as the political control of the orthodoxy remained in force. But this autocratic political patronage was wholly dependent on the whims of the rulers, without any approval of people who were controlled by the orthodox. Thus when the weak Khalifa Mutawakkil wanted to become popular by siding with the increasingly vocal orthodoxy, all hell broke loose against knowledge seekers, first in Baghdad and then in other places, gradually over all parts of the Khelafa.

However, despite the victimisation of scholars like Kindi and Razi, many Muslims still continued to seek new knowledge as their religious duty, though no longer from Baghdad. Ibn Sina defied imprisonment and death threats, refusing to give up his thirst for knowledge, followed by equally determined Ibn Rushd. Some scholars, such as Suhrawardy, were executed, while others were banished as heretics. Even Ibn Khaldun, who was opposed to philosophers such as Razi, was castigated for deviating from the orthodoxy.

The orthodox gained power and authority all the while as Muslims drifted backwards. It is often said with pride that Islam does not have any priesthood, but it must be remembered that even without priesthood, Islam has managed successfully to subvert all freethinking and thus prevent progress. The eagerness of the Muslim orthodoxy to castigate every thinker continues unabated to this day (chapter 13).

The ulama concocted an image of Islamic self-sufficiency in which seeking knowledge outside Islam was considered un-Islamic, in spite of hadiths such as: Go even as far as to China to seek knowledge. Any Muslim who disagreed with this world view of self-sufficiency was charged with impiety. As a result Muslims became blind to knowledge and progress, and consequently reached their current status. When the Ottomans conquered Constantinople in 1453, Arabic science died. Technology cannot survive for long without the backing of science. When the Europeans started to apply their science-based technology in ship-building and military hardware, Muslims had no answer to it. While the Ottomans were aware of this increasing technology gap, the other great Muslim powers of that time, such as the Mughals in India, did not even think about it, believing that the world would just go on as it had done in the past, oblivious of encroaching Western military domination founded on science-based technological superiority. They never saw what had gone wrong. Mughal emperor Aurangzeb chastised his teacher for not giving him a proper education, but never bothered to do anything to establish the necessary educational system.

The ulama viewed the decline as a manifestation of God's displeasure with Muslims for their impiety and the remedy they prescribed was the stricter observation of Sharia, which would somehow restore the past glory. It did not, and in fact the decline worsened.

14.3 Political Reforms for a Democratic Infrastructure

We can conclude from the discussion above that (i) Arabic science flourished because the Mutazilites believed it to be their religious duty to seek scientific knowledge, and that (ii) its greatest weakness was its isolation from the religious beliefs of the masses and its total dependence on the whim of despotic unaccountable rulers. Some later political leaders tried to impose reforms from the top, without appreciating that such militaristic imposition does not work in the long run. Among such reformers were Muhammad Ali of Egypt, Reza Khan of Iran and above all Kemal Ataturk of Turkey. Typically none of them paid much attention to freedom of expression, rule of law, individual rights, or to democratic accountability. In fact a more democratic presidential constitution, originally proposed by Reza Khan in his earlier mellowed days, was rejected as un-Islamic by the Iranian clergy in favour of a monarchy. Their reforms did not work and could not have worked. The study of science, among other creative undertakings, cannot be imposed on a society; it has to grow organically through social, cultural and political reforms. A scholar has to have freedom of speech, freedom to seek knowledge and an environment in which individual rights, rule of law, democratic institutions and democratic accountability are cherished. This is the basic democratic infrastructure without which science cannot be nurtured to flourish.

Today most Muslim countries are undemocratic, without much rule of law or human rights for their own citizens, but full of lip service to these ideals. For instance in 1983 the Heads of the Islamic States approved a Draft Declaration of Human Rights in Dhaka (Bangladesh), while political opponents rotted in the Bangladeshi jails, as in the jails of other participating states! Such is the difference between lofty words and reality in Muslim countries.

Dictatorial countries do not generally motivate people to carry out creative work, except on limited objectives in times of national emergencies. Scientific activities require creativity which cannot be enforced but can only be encouraged with incentives in a conducive environment. Greatest breakthroughs in science were made only by the free spirit and burning desire of individuals.

However, it is true that scientific progress can be attained in a totalitarian environment, as was achieved in the former Soviet Union, but such progress is stilted and command-driven, lacking in the solid foundation of the organic participation of the masses, who do not feel or see themselves as stakeholders in that science. While the Soviet Union could send a man into orbit, it could not even produce the right screws for wardrobe hinges or the right plugs for wash basins in Moscow. In contrast in the USA, science and its technological applications have advanced in all directions in unison, partly because of which the Americans were able to overtake Soviet space science rapidly. This implies that a democratic infrastructure with the people as stakeholders is essential for science and technology to flourish organically.

14.4 Religious Reformation of Islam

As discussed in chapter 12, two great Indian reformers Sir Sayyed Ahmad Khan and Syed Ameer Ali advocated the need to produce a science-based Muslim society, but they made no impact on Muslim society, since what they proposed shattered the Muslim self-image of Islamic self-sufficiency based on what has been labelled in the previous chapter as the *Closed World Assumption* (CWA) on knowledge. In the 20th century the *ahl al-Quran* followers advanced the arguments of Sayyed Ahmad Khan and Syed Ameer Ali, and produced some radical ideas to meet the challenge of modernity and science as explained in the previous chapter. These ideas, so far as the development of science was concerned, were also supported by non-*ahl al-Quran* scholars of the recent past as presented here. However, we begin here with the core topic of the nature of truth as expressed in a human language and its interpretation.

There is an essential difference between science and religion. Science is *falsifiable*, that is, scientists make models of reality that explain known facts and predict new facts; and if at any time a new fact is found to contradict a model then that model must be rejected, that is, *falsified*. In that event a new model, explaining all past facts (including those facts that contradicted the old model), must be made and this new model must also predict some new facts. It is the job of a scientist to actively seek new facts and to continually check if the current model is still valid. This is the scientific process. The scientific truth lasts only as long as its model remains valid.

In contrast, religious truth is not falsifiable, as it is meant to uphold the eternal truth. Thus, as facts change, science changes; but religion does not, though its interpretation can and usually does. This is so with religious truth, since (i) human understanding of the divine word can never be perfect (that is, it will always remain limited and incomplete), and (ii) eternal truth can only be deciphered through appropriate interpretation. Much of the current Muslim dogma is based on the so-called divine interpretation of the Quran developed between the 8th and 10th century CE, with the help of some questionable hadiths. That interpretation has produced the current Muslim state of decline, while the earlier Mutazilite interpretations led to the earlier Golden Age of Islam. How does one then decipher the truth contained in a divine message expressed in a human language?

The effect of human language on divine messages has been discussed in the previous chapter from the *ahl al-Quran* perspective. We shall delve into these issues a little further here. A language has a human element in it, and it evolves as did Arabic, and therefore it is never perfect. A language is strongly related to the culture of the human community that uses the language, the way that community sees the world, the way that community thinks, and the way that community conducts itself. Therefore the word of God expressed in the medium of such an imperfect human language is necessarily constrained by that imperfect media. That is why many of the Quranic verses require contextual interpretation, given by human beings who are also imperfect. Let us consider two examples here:

(i) Take the Quranic ayah [55:5], which as discussed in section 10.1 is considered to be allegorical. It conjures up at the literal level an image of everything physically prostrating before God, while at an abstract level it can be used to deduce a deeper truth of universal laws of God, as claimed by the Mutazilites.
(ii) Take the ayah [33:88] which states: God "protects (all), but is not protected (of any)". This notion of protection makes perfect literal sense in the context of the pre-Islamic Arabic tribal custom, but less clear outside that context; and yet one can perceive in it an iconic truth independent of that social custom. Such an iconic truth is considered to be higher than the literal truth expressed within it.

In general one can make the following observations regarding divine truth and its expression in the human medium, which is necessarily imperfect [Wat88/p88]:

(1) Truth about God cannot be fully comprehended by the human mind and cannot be fully expressed in any human language. Any divine expressions in any language may not necessarily hold literal truth, but they can be allegorical as stated in the Quran itself (section 10.1).
(2) A human element is necessarily present in the divine messages contained in the scripture.
(3) Historical and literary criticism must be accepted, with a view to reaching a deeper meaning, rejecting that which could be an elaboration of allegorical truth.
(4) Scientific methodology for critical analysis in order to distinguish the truth should be accepted.

This view of religious truth is accepted by both *ahl al-Quran* and non *ahl al-Quran* scholars. The *ahl al-Quran* position is to seek new meaning of a divine world as appropriate to a culture at a given time, as discussed in the last chapter. Two well-known non *ahl al-Quran* scholars cited below are the Iranian Shia modernist Ali Shariati[2] (1933-77), and the Pakistani Sunni scholar Fazlur Rahman (see footnote 8 in chapter 13), both of whom subscribed to the above view.

According to Shariati [Ruth91/p347]: ... "the language of religions, particularly the language of the Semitic religions in whose prophets we believe is a symbolic language" – "a multi-faceted language, each aspect of which addresses itself to a particular generation and class of men". ... ..."the ayas of the Quran are allegorical, the sign of God revealed in nature. ... the Quranic view of nature is "closer to the approach of modern science than to that of ancient mysticism".

According to Fazlur Rahman [Rah79/p234]: ".. the Quran, although it is the eternal Word of God, was nevertheless, immediately addressing a given society with a specific social structure. This society could, legally speaking, be made to go so far and no more". Therefore there is nothing wrong in borrowing from the Western cul-

---

[2] Iranian modernist Dr Ali Shariati (1933-77): With a PhD in sociology from Paris in early 1960's, this well-respected intellectual devoted himself to Islamic socialism in Iran, but was harassed and placed under house-arrest by the SAVAK (the Shah's secret police) in Iran, before he was allowed to leave for England where he died suddenly of heart-attack in 1977 at the age of 44 – poisoned by the SAVAK, according to his supporters. He has influenced modernist thinking in Iran, which has been rejected by the current ruling conservatives.

ture. All civilisations grow and in the process borrow from other civilisations, as Muslims did in the past and the Europeans did from the Muslims in turn. ".. Islam did not merely 'borrow', it Islamised all that is borrowed and integrated it into an Islamic framework of values, which in turn was expanded if it was not quite adequate, and this interpretative process occurs with every developing culture". Observe that Fazlur Rahman also denounced the orthodox dogma of Islamic self-sufficiency.

For contextual reinterpretation of the Quranic verses to derive religious truth (as advocated above), there is a need to identify those Quranic verses that deal with eternal values, following the proposal of Abdullahi An-Naim of the Sudan (section 13.3). These verses can then be used to define a basis for Muslim society in the 21st century.

As regards hadiths the position is quite clear. Most Muslim modernists (whether they subscribe to the *ahl al-Quran* doctrine or not) are dubious about the genuineness of all the collected hadiths and believe only a very few of them to be genuine. Even though they do not reject hadiths in principle (except the *ahl al-Quran* followers), they do not treat hadiths as the divine interpreter of the Quran. On the contrary, they assert that hadiths should be studied *only for clues to the spirit behind them* in their situational contexts, *not* for direct application.

As stated before, many Egyptian modernists, such as Taha Hussain, also support an enlightened scriptural interpretation appropriate for the current age. These ideas then provide a basis for the religious reformation of Islam, which can create a conducive socio-religious environment for intellectual activities and for the development of science and technology. Summarising the reforms suggested for this reformation:

1. Examination of the Quranic texts for the identification of the eternal verses for study, in order to apply them to create a basis for a modern Muslim society.
2. New rational and enlightened interpretation of the Quran, consistent with scientific truths, to meet the needs of the time, place and the people, recognising deeper truths, allegorical truths and the imperfect nature of human language.
3. Hadiths to be used, only when proven to be really genuine, "for clues to the spirit behind each situational context", but

not for direct applications, nor as the divine interpretation of the Quran.

In addition, Muslims need to change their attitude towards knowledge gathering from the *Closed World Assumption* (CWA) to the *Open World Assumption* (OWA). They should also accept, as did the Mutazilites, that it is an Islamic imperative to reveal the mystery of God's creation, by seeking knowledge from *all* sources. Using these ideas as the core principles, a new Sharia (the term proposed by An-Naim) can be created by modern Islamic scholars. This is however beyond the scope of this book. So far as this book is concerned, it would seem to us that these core principles should yield the necessary religious reformation for science to flourish.

14.5 Socio-religious Education

Today a dichotomy exists in the educational systems all over the Muslim world, including Indonesia, Malaysia, the Indian subcontinent, Iran, much of the Arab world and even Turkey. The Western educated ruling class controls the administration, while the madrasa-educated ulama control the masses and thus society as a whole. It parallels the earlier battle between the Mutazilites and Hanbalite orthodoxy that took place in the Golden Age of Islam.

The traditional religiously educated ulama, with little knowledge of the modern world, are urging Muslims to go backwards in order to regain their past glory, while the Western educated elite, with little knowledge of Islam, are unable to motivate the masses to pursue modernisation. Western secular law in Turkey left the masses cold and isolated in the absence of any support from the ulama, a problem that was belatedly recognised and addressed in Turkey with the resumption of religious training. In contrast, Saudi Arabia is wholly dominated by the Wahhabi orthodoxy, enthusiastically condemning all forms of modernisation as *bida'*.

For scientific endeavour to flourish, an enlightened (as against conservative) theological perspective as outlined above, has to replace the backward-looking Sharia that dominates Muslim thinking today. As stated earlier, this cannot be imposed from the top, but has to be achieved through education, not for the elite alone, but for all, particularly for the masses, the latter through the madrasas and the ulama. In other words, it is only the ulama who possess the religious authority over the masses to make this religious reformation possible, and therefore it is they who need to be educated in this enlightened and forward-looking version of Islam.

The idea of the Islamic *Closed World Assumption* encouraged the ulama of the Ottoman Empire and elsewhere to shut themselves off from the rest of the world, thereby preventing themselves from learning anything new. For centuries, they remained totally ignorant of the Western thought, and developments in philosophy, science and technology. When suddenly faced with these in the modern world, today's ulama (of the same old tradition) also do not know how to respond, except to a call for more Sharia. They ride a car that has only a reverse gear – they can never move forward. They are unable to accept any historical criticism; instead they hold an exaggerated view of earlier social successes of Muslims (from a selective history) and the evils in Western society, without any attempt to understand and learn from history through an objective analysis. Their traditional self-image of Islamic self-sufficiency makes it hard for them to adjust adequately to the modern world. Historically the madrasas are aimed at providing knowledge of a system of ancient ideas, but *not* at imparting knowledge regarding newer ideas, *nor* at offering knowledge on how to create newer systems. Consequently the madrasas are not interested in inculcating a spirit of enquiry and of independent thought. Therefore reform must begin with madrasa.

The madrasa education system needs to address all these, along with the study of mathematics, logic, philosophy and science, particularly at a higher level. While primary education will instil a spirit of openness and forward thinking, it is at higher education where Islam and modern intellectualism must be integrated for the reformation of Islam. This can only be possible through a programme of reformation undertaken by governments or by other competent authorities. Today fortunately there are some reform-minded ulama, but they do not see the way forward. It is possible for a popular government to push reforms through madrasas with the help of enlightened ulama, as had been carried out successfully in Bangladesh regarding birth control, which is now the Government policy backed by the ulama and the masses[3]. Admittedly birth-control is a relatively triv-

---

[3] It seems a part-Saudi-funded political party in the recent Government is preventing reform these days. Bangladesh, which depends on Saudi oil-aid, has been unable to ban Saudi political funding.

ial issue compared to religious reformation, but it nevertheless implies that the door of the ulama is not entirely closed to reform.

Another associated issue is the practice of rote or passive learning in a madrasa, where in addition, students are not allowed to question or to challenge ideas. Such challenges are considered highly irreverent, not only by madrasas, but also by society. But madrasas enforce this anti-questioning policy more strictly, to the extent of expelling students unless they repent and apologise immediately[4]. The best form of learning is active learning, which cannot be achieved without questioning and challenging. Creativity – intellectual, scientific or otherwise – requires a questioning spirit.

Unless Muslims develop a conducive environment which encourages questioning on any issue (as in Judaism and Christianity), they will not achieve any great intellectual breakthroughs in science, or in other branches of knowledge. Obviously, once questioning is allowed, it will not stop at science, it will spread over into religious territories as well. Instead of being frightened, one must welcome such questioning as a means of strengthening one's faith. A faith that cannot stand questioning cannot claim to be a sound faith.

14.6 Way Forward Towards Revival

We have discussed the infrastructure necessary for major developments of science and technology in Muslim societies. It is science we are after, technology is assumed to be a spin off from the former. We are aware that any Muslim today in any country can take up science, in spite of the current religious and political constraints. Therefore one cannot say that scientific activities cannot take place in the absence of political and religious freedom. But what one can say and we are saying is that if we wish to see an accelerated and flourishing development of science in a Muslim society, then science has to be integrated into the fabric of that society, in its thoughts, hopes and aspirations, so that the society becomes a stakeholder in such a development. This integration can be best achieved, if scientific activities are viewed as a religious duty in terms of the Quran and hadiths,

---

[4] For example, I know a Palestinian, now working in a UN agency, who was expelled from his semi-religious school for questioning the condemnation of his poor parents by his teacher in the classroom (for borrowing money with interest for his education). He was lucky to get an education after that expulsion.

as was viewed by the Mutazilites. That is why the role of madrasas is so important.

An associated issue that concerns many Muslims are the limits to the scope for scientific inquiry. Here a change of attitude is necessary. Some Muslims seem to support only utilitarian science, forgetting that such science could not have produced Newton's theory of gravity, Maxwell's theory of electro-magnetism, Einstein's theory of relativity or Heisenberg's theory of quantum uncertainty, theories that underpin modern technological civilisation – from electricity to everything. Professor A. Salam (the late Nobel Physicist and a devout Muslim) whose seminal work on *Weak Interaction* has challenged our fundamental ideas about the nature of the universe, was totally opposed to such utilitarian science. Indeed it is the restriction imposed by utilitarianism that discouraged even the study of algebra in Muslim lands (they could not see a use for it). This utilitarian approach has driven Muslims to this current state. Following the Greeks, pursuit of knowledge should be regarded as a virtue in itself, even though society should have the right to limit the scope of its application in some areas, such as on ethical grounds, concerning say weapons of mass destruction or human cloning. In any case, given the backwardness of Muslims in science and technology, this debate is not the most pressing one today. If Muslim science takes off, there will be plenty of opportunities to revisit and debate these issues at that time.

Whether Muslim countries will rise up to the challenge and undertake the necessary reform is a moot point. As we have shown in previous chapters, Muslim states, after the Abbasid Golden Age took very little interest in science. More recently Muslim countries are saying much, but are doing little. Nevertheless, words are better than silence, since words at least express an interest, and thus keep the hope alive that one day words will be matched by deeds.

The position discussed above refers to Muslim majority countries. However, the situation is different in Muslim *minority* countries even though there is a common pattern of disincentive due to religious orthodoxy. In this respect, we shall cite India and then mention some Western countries. India is advancing fast in technology, but the participation of Indian Muslims in that sector is far lower than their population size would suggest. The relative backwardness of Indian Muslims compared to Hindus in the take up of Western education is well-known. It is something Sir Sayyed Ahmad

Khan and others reformers tried to reverse in the face of opposition from the orthodoxy. Today there are of course many Western-educated Muslims in India, but they represent a far lower fraction of the Muslim population than do the similarly educated Hindus in the Hindu population. Furthermore, on science, the Muslim fraction is even further lower than the Hindu fraction. Fatwas, such as the Internet being haram, could not possibly have helped Muslims to enhance their interest in science. The relative backwardness of Pakistan in science (despite its Bomb technology) compared to India has the same root cause. Obviously religious reformation of Islam as discussed here would help Indian Muslims, whereas in case of Pakistan there is an additional requirement for democratic institutions. As for the other Muslim countries, it seems Malaysia and Indonesia could be the ones that might in the future provide conducive environments for scientific revival.

While we have discussed above countries with majority or large Muslim populations, we cannot ignore Muslims in Western countries, such as in Britain, France, Germany and the USA. In fact it may be that religious reformation could emanate from them. In this connection it should be noted that the French Government is keen to build a French Islam (with loyalty to the French constitution) as it did in the 19th century in order to create French Jews. A brief digression on this theme is perhaps helpful.

Judaism suffered from the orthodoxy in the middle ages as do Muslims today. They looked backward, shunning all modernisation. A major reformer was Moses Mendelssohn (1729-86) of Germany who insisted that reason preceded faith. He is said to have privatised Judaism by separating it from the state. Although this doctrine was an anathema to the orthodox at that time, it liberated Jews from the ghetto culture and permitted them to participate in the idealism of *Enlightenment* that characterised that age. Freed from the yoke of orthodoxy, Jews started making major contributions to science and the secular activities of the state. Encouraged by Napoleon, French Jews declared the French Revolution as the "second law from Mount Sinai" in 1806 and two years later they accepted, what some say a "Faustian bargain" of, full-fledged French citizenship free from harassment, in return for the privatisation of Judaism. This French solution became the pattern for the rest of the Europe. Jewish people thus became French, German, or English as the case may be, rather than remaining segregated ghettoised European Jews. Jewish contributions to science, technology, medicine and other forms of intellectual pursuit began to flourish.

It would seem that the French authorities today are planning a similar type of reform in order to create French Muslims, loyal to the French state, with the support of some French ulama. A programme of the training imams for some 900 French mosques has already begun. If this experiment is successful, then this may show a way forward for Muslims in other Western countries. It is unlikely that the British or US Governments will *directly* enforce any such reforms, but the British Government will probably provide assistance for reform programmes.

14.7 Conclusion

In some sense the religious reformation, suggested here is meant to provide the *scope,* and the *capacity*, within a democratic infrastructure for scientific activities in Muslim society, borrowing the two terms from Fukuyama [Fuk04]. The third important element that will help speedy progress is religious *motivation*, which is in fact provided in Islam by the exhortation in the Quran and hadiths to understand the mystery of God's creation. This exhortation should make the study of science and technology a religious duty for Muslims, as was accepted by the Mutazilites in the Abbasid Golden Age. Strictly speaking this third element is really part of religious reformation with emphasis on science and technology, although we have cited it as a separate item in order to underscore its importance. It seems to us that these three elements, apart from opportunities in terms of funding and facilities, could provide a sound foundation for the creation of a science-based society under Islam. However, the reform of the madrasa-education for the ulama is essential for the building of mass-support for these ideals, without which science will not take root in Muslim society. Clearly all these major reforms would take generations to emerge. And even then it would be a long slow process, because of the time and devotion needed for the study of science. As George Sarton said [Sart50/p42]:

> "There is no short road to science. It is easy to instruct mechanics, and not very difficult to train engineers, but the education of a man of science is a long and arduous process. The way to science is hard without rest and endless. Creative

achievements in science or learning require a man's uninterrupted devotion for many years if not for a life-time".

Finally we have tried in this book to discuss the rise and decline of Arabic science with a view to analysing the contributory factors. Based on this analysis, we have also suggested what Muslims need to do to revive Muslim science, noting that by Muslim science we mean Muslim contributions to science, science being both secular and international. The great thing about the Western civilisation is that it is open and it samples from all sources for its advancement and vitality, as Muslims did in their Golden Age. However, unless Muslims rise to the challenge and embark on a programme of scientific development, they will remain backward and marginalised in the increasingly science-driven technological world of today. The answer to the question whether they will rise to the challenge should best be left to the Quran:

> "Verily God does not change the state of a people unless they change the state of their inner selves." Quran [13:11].

It is important to make a start, sooner rather than later, on the long road to a science-based Muslim society, supported by an Islamic reformation movement, which includes forward-looking reformist scholars and lay people, who believe in this imperative.

References and Sources

[Ann96] A. A. An-Naim: Towards Islamic Reformation, Syracuse University Press, 1996 (pbk). [See the same reference in chapter 13 for more information on this work].

[Bro99]: Daniel Brown: Rethinking Tradition in Modern Islamic Thought, Cambridge University Press, 1999.

[Fuk04]: Francis Fukuyama: State Building: Government and World Order in the Twenty-first Century, Profile Books, 2004.

[Hoo91] P. Hoodbhoy: Science and Islam, Zed Books, 1991.

[Hou83] Albert Hourani: Arabic Thought in the Liberal Age (1798-1939), Cambridge University Press, 1983.

[Huf95] T. E. Huff: The Rise of Early Modern Science, Chapter 2, Cambridge University Press, pbk edition 1995.

[Rah79] Fazlur Rahman: Islam, University of Chicago Press, 2nd ed. 1979.

[Rah84] Fazlur Rahman: Islamic Methodology in History, Islamic Research Institute, Pakistan, [1965, 1984].

[Rut91] M. Ruthven: Islam in the World, Penguin 1991.

[Sard77] Ziauddin Sardar: Science, Technology and Development in the Muslim world, 1977.

[Sard82] Ziauddin Sardar: Science and Technology in the Middle East – A guide to Issues, ... , Longman, 1982. The book gives science & technology profiles of the Middle Eastern countries, including Pakistan, Iran and Turkey, up to about 1980.

[Sart50] George Sarton: The incubation of Western Culture in the Middle East" – a lecture delivered in March 1950 at the Library of Congress in Washington DC, Library of Congress No: 51-60324.

[Sart27] George Sarton: Introduction to the History of Science, Carnegie Institute of Washington, 1927. He has 3 huge volumes (I, II, III) on Science from Homer to the 14th Century, some 5000 pages. He also has two further volumes with the title "A History of Science" on ancient and Greek sciences cited above.

[Wat87] W. M. Watt: Islamic Political Thought, Edinburgh Paperback Edinburgh University Press, 1987.

[Wat88] W. M. Watt: Islamic Fundamentalism and Modernity, Routledge, London 1988.

[Zak89] R. Zakaria, The Struggle Within Islam, Penguin, 1989.

GLOSSARY

All terms used in the book have been explained in the main text at the appropriate locations which can be found from the Index given after this Glossary. In this Glossary some of the more commonly used terms have been restated – in some cases with chapter/section or page number in the main text.

Abbasid Dynasty
The Abbasid dynasty ruled Islam, with their capital mostly in Baghdad, from 750 CE until 1258 CE, when Baghdad was sacked and the last Khalifa brutally killed by the Mongol Halagu Khan. The Abbasid dynasty created the Golden Age of Islam within its first 100 years of its reign. See section 2.4 for details.

Abu Hanifa
Imam Abu Hanifa (699-767 CE) is revered as the architect of the Kufa school of law (madhab) of Islam. He produced a systematic theoretical approach to technical legal thoughts, and hence the Kufa school was later named by his followers as the Hanafi school. Abu Hanifa, a former merchant from Kufa and a convert, pioneered the new discipline of jurisprudence (fiqh). He did not subscribe to the *as'hab al-hadith* (the partisans of the hadiths) doctrine, as he believed in *ijtihad* (independent reasoning) with the ability to make new laws.

Abu Huraira
Abu Huraira, a Companion of the Prophet, became Muslim in AH 7. Despite the companionship of only 2½ years with the Prophet (who died in AH 10), Abu Huraira is the source of the greatest number of hadiths (5376), compared to 2210 of Ayesha and 132 of Abu Bakr – see pp218/219). He never wrote any hadith down, but claimed to have a perfect memory. His hadiths are controversial (chapter 13).

ahl al-bayt
It means people of the [Prophet's] household, which in the Prophet's lifetime included Ali (Prophet's cousin and adopted son – the 4th Khalifa), Ali's wife Fatima (Prophet's daughter) and the Prophet's grand children Hasan and Hussain. Later the descendants of Hasan and Hussain were regarded as members of *ahl al-Bayt*. The Shias accept only the members of *ahl al-Bayt* to have the right to rule Islam.

ahl al-hadith
An 18th century Islamic purification movement, partly linked to the Saudi Wahhabi sect, gave rise to an *ahl al-hadith* (people of the hadith) group of reformers who believed in a single literal interpretation of the texts of the Quran, provided through hadiths and only through hadiths. They called this interpretation as the divine interpretation, which in their view, provided salvation for Muslims. They were interested to re-study hadiths and also to re-examine their authenticity (chapters 13 and 14).

ahl al-Quran
The trauma of the Muslims in the 19th century at the hands of the Western Imperialists resulted in a drastic reform movement in the early 20th century, both in India and in Egypt. These new reformers did not consider the *ahl al-hadith* doctrine to be radical enough to meet the new challenges of science, technology and Western dominance. They produced a new doctrine of *ahl al-Quran* (people of the Quran) in which the Quran alone was perceived to be self-sufficient for Muslims, to the exclusion of hadiths, which they argued were unreliable anyway (chapters 13 and 14). They believed in the human interpretation of the Quran, but **not** on any divine interpretation through hadiths. Their views are closer to the Mutazilite doctrine (chapter 3).

Alim: Singular of ulama, see Ulama

Amirul Mu'mineen
Khalifa Umar (the second Khalifa) believed the title Khalifa (successor [to the Prophet]) to be too grandiose for him, and hence he accepted the title *Amirul Mu'mineen* (Commander of the Faithful) as a humbler alternative. However, the later Khalifas employed *Amirul Mu'mineen* as a grand title to emphasise their worldly power.

Arabic Science
Arabic Science is the universal science created under Islam by Arabs, Non-Arabs, Muslims and Non-Muslims. They all used the Arabic language to discuss and document that science, and hence it is Arabic science.

As'hab al-hadith
A pro-hadith group called *as'hab al-hadith* (the partisans of the tradition) came into existence in the second century of Islam in around 770 CE (some 120 years after the death of the Prophet), demanding all laws to be based on explicit hadiths. The schools of Kufa and Medina initially resisted the pro-hadith approach of the partisans, but had to give in later, even though not completely, when at the end of the second century Imam Shafi'i (767-820) elevated the pro-hadith doctrine onto a level where hadiths were the only valid authority for interpreting the Quran (chapters 4 and 13).

Astrolabe
The astrolabes were the most treasured and sophisticated astronomical instruments in the middle ages, which displayed a mathematical model of the heavenly bodies.

GLOSSARY

Astrolabes were used (before the invention of sextant) to observe the position of celestial bodies (section 6.3).

Ayahs: Verses in the Quran

Ayesha
The favourite wife of the Prophet and a source of hadiths and sunna (tradition). She died in 678 CE, aged 64. She was the daughter of Abu Bakr – the first Khalifa.

Bida'
Bida' is assumed to mean innovation and later interpreted to imply anything outside custom and practices legitimised by Sharia. In its extreme it advocates the rejection of all new knowledge, ideas and amenities that were not known to, commented upon or enjoyed by the Prophet, which is in fact everything that the modern civilisation offers. It was used to stifle science and philosophy. Later on, *bida'* was divided into good and bad *bida'*s, only the bad ones to be forbidden – the bad *bida'*s were the ones that were contrary to the Quran, hadiths and *ijma'*. However, since *ijma'* depends on the climate of opinion which may be based on needs and may vary from place to place and time to time, today's *bida'* can become the tomorrow's sunna (tradition).

Bukhari
The most famous hadith compiler was Mohammad bin Ismail al-Bukhari (810-870). His collection, often referred as al-Bukhari (or just Bukhari) contains around 7,275 *sahih* (as defined by al-Bukhari) hadiths which includes some 3000 repeats, as the same hadith is sometimes repeated under different subject-categories [pp59/60]. The al-Bukhari hadiths are considered by most Muslims as most authentic. See also the entry on Hadiths.

Existents: Things in existence including living things.

Ex-nihilo: Out of nothing, e.g. the creation of the universe out of nothing.

Falsafa (or falsafah): Islamic philosophy based on Greek thought.

Fatwa
A fatwa (a legal opinion) can be issued by any Muslim scholar, and therefore it is always possible to find a favourable scholar to issue a desired fatwa. Once someone is denounced by a fatwa as a heretic, no amount of counter fatwas can remove entirely the suspicion from the minds of the masses.

Hadiths
Hadiths are the sayings of the Prophet. While a hadith of the Prophet is his saying, there can be hadiths of other people as well. Of particular interest to us are the hadiths of his important Companions related to what the Prophet said or did, and also concerning what he approved or disapproved of. A hadith, which was originally meant to be an oral report, has two parts: *isnad* (transmission-chain) and *matn* (text). The *isnad* describes the successive transmitters of the *matn*, linking

the current transmitter to the source (which in the case of the Prophet is one of his Companions). For example:

> *Ibn Ishaq reports from Sa'id bin Abu Sa'id from Abu Huraira* that the Prophet said " ......".

The words in the italic constitute the *isnad*, and what the Prophet actually said (not shown above) is the *matn* (text). The earlier hadiths did not have *isnads*. These were added later, starting from the turn of the 1st century AH (7th/8th century CE). In this book, a hadith means a hadith of the Prophet, unless otherwise qualified. The *isnads* of these hadiths go only up to the Companions who are believed to be telling the truth. However, while the Quran had been most carefully preserved from the beginning, hadiths had not. It is not easy to separate sound hadiths from fake hadiths – some Muslim scholars distrust hadiths even in the Bukhari and Muslim collections (chapter 13).

Hanafi: See madhab

Houris (in paradise): Beautiful maidens who will serve the believers in the paradise.

Ijma'
Ijma' (consensus) is the third source of Sharia law (after the Quran and hadiths). Imam Shafi'i interpreted it as the consensus of all Muslim ulama, rather than that of the ulama of a locality, as hitherto had been interpreted by the Hanafi and the Maliki schools. *Ijma'* cannot contradict the Quran or hadiths.

Ijtihad
Ijtihad is the fourth source of law in Sharia, and it was defined as personal intellectual struggle. A particular form of *ijtihad* is the reasoning by analogy or *qiyas*. *Ijtihad* can be applied only when it does not violate the other three sources (the Quran, hadiths and *ijma'*), but subsequently the scope of *ijtihad* had been narrowed down, first only to *qiyas* and then only to a special form of *qiyas*, devoid of any *ra'y* (considered personal opinion), especially denuded of any personal discretion (*istihsan*). Although no one formally declared it, the gate of *ijtihad* was deemed closed in the 10th century, and the right of *ijtihad* was replaced by a new doctrine of the duty of *taqlid* (imitation). Henceforth every jurist was an imitator, bound to accept and follow the doctrine of the predecessors, rather than taking recourse to *ijtihad* to produce a new solution.

Imam
Imam means leader. There are many kinds of leader, such as the Imams of the Shia Islam (e.g. 4th Khalifa Ali and some of his descendants), a Shia leader (e.g. Imam Khomeini of Iran), Imam of a school of law (e.g. Imam Abu Hanifa), Imam of a mosque (one who is appointed to lead the regular prayers). Any Muslim (even if he is not appointed as an Imam of a mosque) can also lead a regular prayer in that mosque (if the Muslims participating in that particular prayer agree) – in this case that leader is the Imam of that particular prayer.

Imam Shafi'i

Imam Muhammad Abu Idris Shafi'i (767-820), a legal genius, produced a legal theory purely based on the Quran and hadiths (independent of the local differences) as the foundation of Sharia. His followers created the Shafi'i school (madhab) of law.

isnad: See hadiths.

Janissaries

Ottoman sultan Murad I (1362-89) created a unit of bodyguards, later called *Janissaries* (meaning new armies), for the protection of the sultan. The Janissaries, like the Mamluks (section 2.6), were slaves recruited as boys, converted into Islam and trained for the army, but unlike the Mamluks they never became rulers themselves, although they subsequently played a major role in managing the Ottoman state.

Kafir: A kafir is an unbeliever, also referred to in this book and elsewhere as a heretic or infidel (see also page xiii).

Kalam

The Asharite theologians develop a compromise doctrine between the Mutazilites and the traditionalists, based on reason and logic, in which God is beyond our understanding. However they were fundamentally opposed, as were the traditionalists, to science. Their doctrine later led to the philosophy of *ilm-al kalam* (knowledge of kalam, or just *kalam*), literally knowledge of discourse. The term *kalam* is sometimes interpreted to mean theology, and hence the Mutazilite doctrine is also described sometimes as the Mutazilite *kalam*.

Khalifa

Khalifa (or Caliph its anglicise corruption) means successor. It was used as the title by the Rashidun, and also by the Umayyad and Abbasid rulers, to mean Successor [to the Prophet]. Some other Muslim rulers also called themselves Khalifa. Ottoman Sultan Abdul Hamid II (1876-1909) revived this title to project himself as the leader of all Muslims, including even those outside the Ottoman Empire (see section 11.2).

Khelafa: Its anglicised corruption is Caliphate – the empire ruled by a Khalifa.

Khutba

A Khutba is the obligatory sermon delivered by the Imam in the mosque after the Friday Jumma (mid day) prayer. In it the name of the legitimate ruler of the day is mentioned, and hence Friday Khutbas were watched by the new Muslim rulers.

Madhabs

Madhabs are the Islamic schools of law (or theology) under Sunni Islam. There are four important madhabs: Hanafi, Maliki, Shafi'i and Hanbali. Each madhab regards the other madhabs as valid. The Hanafi school (developed in Kufa) was more liberal and broad-minded, and was particularly favoured by the Abbasids, the Ottomans and the Mughals. The Maliki school was developed from the ideas of

Malik Ibn Anas (715-95) in Media, and therefore in comparison with the Hanafi school, it was narrower in its outlook reflecting the relative conservatism of Medina. Malik's two distinguished pupils were Shabyani who joined the Hanafi school and Shafi'i whose followers established the Shafi'i school, based on his legal theory and ideas. The conservative Imam Hanbal created the Hanbali school, which nearly died out in the 14th century CE, until it was revived by the ultra-conservative Muhammad Ibn Abd al-Wahhab (d 1787). Because of their intolerant attitude to fellow Muslims, the earlier Wahhabis were disliked, until the recent times when the oil-rich Saudi Arabia (Guardian of the two Muslim holy places) adopted Hanbalism-Wahhabism as its official madhab.

Madrasa: Religious schools.

Mamluks
Mamluks were a crack military corps composed of Circassian slaves captured as boys and converted to Islam, similar to the Janissaries of the Ottomans, but unlike the Janissaries, the Mamluks ruled over their country, viz. Egypt. They started their rule of Egypt in 1250 and continued in one form or another until they were eliminated by Muhammad Ali of Egypt in 1805.

Muslim (hadiths)
The second most revered hadith compiler (al-Bukhari is the first) is al-Muslim bin Hajjaj (817-874 CE). His collection contains about 4,000 *sahih* (as defined by al-Muslim) hadiths. However, many hadiths in the two collections (al-Bukhari and al-Muslim) are common, but al-Bukhari is usually regarded as more authentic, since he had applied stricter acceptance criteria. See also the entry on Hadiths.

Neoplatonism
Some 600 years after Aristotle, Plotinus, who lived first in Alexandria and then in Rome, formulated the new philosophy of Neoplatonism, re-interpreting the ideas of Plato on God and creation. In it God is referred to as the *One*, the source of all existence. The *One* is totally other than all else, transcending Aristotle's concept of God. Furthermore the *One* is not only the first cause (which keeps the world in perpetual motion), but the *One* is also the source from which all existents emanate, like the light emanating from the sun. Emanation is an eternal creative process. Plotinus also subscribed to the concept of *soul* that controls the body. Many Muslim philosophers found Neoplatonism as an attractive base for the development of their version of Islamic philosophy (see chapter 10).

Rashidun
The first four Khalifas (viz. Abu Bakr, Umar, Uthman and Ali) after the Prophet were known as the *Rashidun* – the rightly guided ones, who used the state to advance the cause of Islam and Muslims, while the later Khalifas employed the state to serve themselves and their egos. While the first four lived simple lives with meagre sustenance from the state in austere conditions (except for Uthman who had previous personal wealth), the later ones lived in palaces, in opulence and in luxury, paid for by the state. The first four Khalifas were addressed directly by their names. Umar thought the title *Khalifa* [to the Prophet] as too grandiose for

him, and therefore he called himself *Khalifa to the Khalifa* [to the Prophet], and also accepted for himself what he considered to be a humbler title: *Amirul Mu'mineen* (Commander of the Faithful).

Sheikh al-Islam
Sheikh al-Islam was the third highest person (after the Sultan and the Vizier) under the Ottomans, as the head of religious affairs. He would issue the fatwas for the state. Following the Ottomans, the Qajar rulers in Shia Iran also created this post.

Sharia
Sharia or Sharia law is the Islamic law produced by the ulama between the 9th and 11th century. The ulama believed that only sunna (the apostolic tradition) and hadiths (sayings of the Prophet), but not human intellect, provided the valid (in fact the divine) interpretation of the Quran, and hence the basis of Islamic law. The resultant Sharia law emcompasses not only civil and criminal laws, but also laws on religious matters, such as, religious rites, rules, duties and obligations. By shaping the everyday life of Muslims, Sharia has also created the Muslim attitudes towards science and knowledge (chapter 4).

Shia and Sunni
Shias and Sunnis are the two main branches of Islam. The Sunnis include some 90 percents of the Muslims in the world. Both believe in the Quran and the Prophet, but the Shias do not recognise the first three Khalifas of Islam (namely Abu Bakr, Umar and Uthman) as legitimate successors. They accept only Ali (the 4th Khalifa – the first Shia Imam) as the legitimate Khalifa. Therefore while the Sunnis accept the religious opinions and edicts of all the four Khalifas as valid, the Shias accept only those of Ali as valid. While the Shias belong to *shi'at al-Ali* (party of Ali), the Sunnis belong to *ahl al-Sunna* (people of the tradition) of the Prophet. However during the hajj, both Shias and Sunnis usually pray together in Mecca under the same prayer leader (Imam), who these days belongs to the Saudi Wahhabi sect (i.e. madhab).

Sunna
A sunna (the apostolic tradition) of the Prophet is normally supported by a hadith of the Prophet or of a Companion.

Sunni: See the entry for Shia.

Taqlid: See ijtihad.

Ulama
Ulama (sing. alim) means religious scholars, whose education is usually limited to only the religious subjects, such as the study of the Quran, hadiths and Islamic law (Sharia). They do not study science or philosophy, not even arithmetic beyond what is needed for the calculation of Muslim property inheritance (according to Sharia). They are usually very conservative, and do not study other religions, nor any modern subjects. They usually receive their education in madrasas (Islamic religious schools).

Umayyads
The first dynasty of Muslim rulers, established by Muawiya in 661 CE as the 5th Khalifa after the death of Ali. They ruled Islam from Damascus for some 90 years, and expanded the empire from Spain to Sind. They established sound administrative polities, but their rule was unpopular, particularly because of their massacre at Kerbala and also because of their treatment of the non-Arab Muslims (who were the majority) as second class citizens (see section 2.3).

Wahhabi: See madhabs.

Index

This index does not generally include names/items that appear in section and subsection headers within chapters, such as Gazzali or Ibn Sina. It also excludes the names of the Khalifas, kings and emperors (chapters 2 and 11), as they can be found from the obvious (sub)sections. There are some exceptions to this, such as Timur Lang, who is important but does not belong to any obvious subsection, and hence their names are included. Observe also that in some cases, names starting with al, Ibn or Imam are listed below without these prefixes. The numbers given are the page numbers, except where prefixed by *sc* (for section), or *fn* (for footnote), the latter with a page number prefixed by p.

Abbas 16-17, 28
Abbasid *sc* 2.4
Abd al-Malik *sc* 2.3
Abd al-Sallam 122
Abduh 193, 197, 199, 214
abjad numbers 76-77
Abu Daud (hadiths) 60, 215
Abu Hanifa 29, 63, 48, 63-68, 221
Abu Huraira 55, 218-220
Abu Rayya 224
Abu Sufian 16-18
Abu Talib 15-17
Abu Yusuf (Hanafi) 68, 102
Abu Zaid 126, 217, 225
ahl al-bayt 19,24-5, 27
ahl al-hadith 71, 213-4, 221, 224
ahl al-Quran 71, 147, 214, 221- 29, 238-41
Akbar Ahmad 122
Akbar the Great *sc*11.1
Alexander the Great 10-11
Al-Hijra (AH) xiv, 17
Ali Shariati 240
amirul mu'mineen 22, 28
Anaximander 9
Anaxogorus – the father of Athenian science 9-10

An-Naim 227
Anushirwan (emperor) 12
Ardashir (emperor) 11
Aristarchus of Samos – the Copernicus of Antiquity 10
Aristotle 4, 7-10
asabiya (of Khaldun) 158-59
as'hab al-hadith 68, 215-17, 223
Ashari *sc* 3.3
Ashura 25
astrolabe *sc* 6.3
Athenian Academy 11
Atiqah (Umayyad Queen) 26
Aurangzeb 141, 169, 172-4, 217, 236
Ayesha 19-21, 219

Bactria 11
Badr (battle) 16-17
Baibars 33
Barmakid 29, 105
Batani 88
batin meaning 146
bida' 127
Biruni 93, 98, 104, 106, 126, 138, 150, 172
Bukhari (hadiths) 59-62, 108, 218-20, 223

INDEX

Buyids 32-33, 39, 153
celestial globe 99
Chakralawi 214
Chengis Khan xiv, 32, 96
clock 185-86
closed world assumption 199, 238, 242-43
Constantine 11
Copernicus model 97
Council of Nicaea 11
crescent visibility 93

Darul Funun 197
decline of the Abbasids 32
Democritus 10
destruction of Baghdad 32
Diodatus 11
divine attributes 44-45
divine justice 45

East India Company 170-71
economy (Mughal) 174-5
Edessa 12
falsafa 145-47
faqih 68
Fatima 18-21
Fatimids 39
fatwa p124/*fn*2
Fazlur Rahman *ch*13/*fn*8
fiqh 68
freewill 45
Galen 11, 29
Galenic System 108, *sc*8.1
Golden Age 27-29
Ghulam Pervez 224
Gulbadan 167

hadiths
  fabrication 55-57
  unreliability *sc*13.2
  validation and verification 56
  on seeking knowledge 4-5
Hafsa 19, 21, 23
Hakam-II 35
Halagu Khan 96, 100
Hamza 16-18
Hanbali (madhab) 68-9, 121-4, 213, 218, 242

Hasan Basri 43, 58, 222
Hasan Sabbah 32
Hassan Hanafi 217
Heracleitus 10
Hippocrates of Cos 10
Hisham 26, 28
House of wisdom (Baitul Hikmah) 5
Hudaibiya, treaty of 17

Ibn Bajja 35
Ibn Hazm 35
Ibn Khatib 112
Ibn Masud p215/*fn*1
Ibn Nafis 111, 138
Ibn Salah 123-4
Ibn Shatir 96-99, 130, 138, 140-41, 232
Ibn Taymiyya 124-27, 131, 161
Ibn Tufayl 35
Ibn Yunus 105
Ibn Zuhr 35
Idrisi 104
ijma 64, 67-70, 127
ijtihad 67-70, 122, 186, 197, 211-13
Imam Hanbal 31, 49-51, 56, 60-61, 67-9, 121-2, 141, 219
Ionia, Ionian, 9
Islamic Architecture –birth of, 26
isnad 54-62, 69, 136
Itimad ad Dowla 168-69

janissaries 39, 176, 180, 200
Jawhari 82
Jerusalem 22, 25-26/*fn*5, 55/*fn*3
Jundishapur 12
Justinian 11
Jyrajpuri 224
kalam 49, 124, 139, 216
Kamal al-Din 103, 123
Kashi 77, 86
Kerbala massacre 24
Khadija 15
Khalid bin Yazid 13
khariji 22
Khazin 85
khedive 204
Khuzistan 107
Khwarizmi 76-82, 88, 104, 129

INDEX

Khwarizm-shah 126
Kösem (Ottoman) 170
Kutbi 105
last Abbasid Khalifa
   Mustasim 32
   Mutawakkil-III 33

madhabs 68-9
madrasa education 235, 243-44
Malik Ibn Anas 29, 59, 63, 66-8. 219
Maliki school (see madhab)
mamluk 33, 39-40, 200
Mamun 5-6, *sc*1.3, 28-31,
   49-50,66,79,88,100,104,121
Mansur 28-30, 66, 75, 107, 134
Maragha model 96-99
Marwan 20-21
Marwazi 94
Masarra 35
matn 54, 56, 59
Maturidi 46
Maududi 221
Mendelsshon 246
mihna 31,50,121
Miletus 9, 10
mongol invasion 32
Monophysites 11
Muawiya 21-25, 55, 131, 218-19
Mughal education 172-3
Mughal science 171
Muhammad bin Abu Bakr 20
Mumtaz Mahal 168
Musa bin Shaker 104
Muslim (hadiths) 59-62
Mutawakkil 31, 118, 121, 149, 235

Nawal al-Saadawi 217
Nayrizi 82
neoplatonism 145-46
Nestorians 11
Nizam al-Mulk 32
Nurbanu (Ottoman) 178
Nurjahan (Mughal) 168
open world assumption 199, 211, 242
Ottoman − Khalifa title 177,181
Ottoman geography 126
Ottoman science 182-85
Pahlavi (dynasty) 38, 206

Pahlavi (lang) 12
paper-making 105, 125, 130, 133
Plato 10
printing 185-86
Ptolemy, model, 89, 94-95
puppet khalifa 33

Qadarires 42, 222
qanun 69, 77
qibla 92
qiyas 67-68
Quraish 14
Quran
   allegorical ayah 44, 148, 239-41
   compilation 23
   createdness 45, 48
   khalq-i-Quran *sc*3.2
   uncreated 45, 48-9, 51, 145

Rashid Rida 224
Rashid (Khalifa) 5, 28-30, 49, 105,
                     109,134, 225
Rashidun 16, *sc*2.2
Raziq 135
religion of science 9
revelation *vs* rational reasoning 146
Roger II 104-05

Sabians of Harran 12
Saladin 26
Salam xi, xii,117-8, 146
Samarra 31
Sardar, Ziauddin 234
Sarton 232
Sasanid dynasty 12, 20, 38
schools of law (see madhabs)
Seljuks 32, 160, 172
Shafi'i 52, 56-63, 66-71, 124, 215-16,
                         223
Shafi'i's legal theory 66-8
Shahjahan 34, 168-72
Shawkani 213
Shaybani (Hanafi) 59, 68
Sheikh al-Islam 178
Shirazi 97
Sidqi 224
Siege of Vienna 178-79
Sind 26

sindhind 75, 88
Sirhindi (India) 167
soap invention 104
Suhrawardy 134, 155
Sulaiman the magnificent xiv, 177-79, 184
Sultan Mahmud 93
sultanate of the women 177-79
sunna 54, 59, 63, 66-69
Susa 12
Syriac 5, 12/*fn*5

tafakkur 4
Taha Hussain p216
Tajmahal – destruction plan 170/*fn*4
takfir 22
tanzimat reform 180-81
taqlid 70
tashkeel 4
Thabit (maths) 82
Thales 9
Theodosius 11
Timur 100, 158, 176
Tirmidhi (hadiths) 61, 215
Trench (battle) 17

Turhan (Ottoman) 176
Tusi 83, 96
Tusi Couple 96-97

Uhud (battle) 17-18
Ulugh Beg 86, 100
Umar Khayyam 32/*fn*11, 85,
Umayyads sc2.3
uqlidisi 77
Urdi 97-98

Wahhabi 69, 213, 125, 127, 242
Wali Allah 197
Walid (Khalifa) 26
Wasil Ata 43
wine-drinking 47-48

Yathrib p17/*fn*1
Yazid 23
zahir meaning 146
Zaid bin Thabit (Quran) 23
Zain al-Abedin 24,25.
Zaqali 35
zij 75
Zubeida 28